Studies in the History of Science

UNIVERSITY OF PENNSYLVANIA
BICENTENNIAL CONFERENCE

Studies in the History of Science

By

E. A. SPEISER

OTTO E. NEUGEBAUER

HERMANN RANKE

HENRY E. SIGERIST

RICHARD H. SHRYOCK

EVARTS A. GRAHAM

EDGAR A. SINGER

HERMANN WEYL

UNIVERSITY OF PENNSYLVANIA PRESS
Philadelphia
1941

Manufactured in the United States of America
by The Haddon Craftsmen, Inc., Camden, N. J.

Contents

UNIVERSITY OF PENNSYLVANIA
BICENTENNIAL CONFERENCE

Ancient Mesopotamia and the Beginnings of Science

By

E. A. SPEISER, PH.D.*

THE thesis which this paper aims to outline embodies the following propositions: (1) Available evidence points to Mesopotamia as the oldest known center of scientific observation permanently recorded. (2) Whatever its immediate objectives, this activity comes to include such widely separated fields as education and language study, jurisprudence, and the mathematical and natural sciences. (3) The numerous elements in this broad advance are interrelated basically. The common underlying factor to which the initial impetus can be traced is a concept of society whereby the powers of the state are restricted and the rights of the individual receive a corresponding emphasis. (4) It is significant that under the opposite social system of totalitarian Egypt early scientific development differed in scope as well as in degree; while notable in certain special fields, such as medicine and engineering, it lacks the breadth and balance manifested in contemporary Mesopotamia.

At this point it is in order to insert a remark of explanation. Although the present paper is listed under Natural Sciences, its specifically scientific content is negligible; furthermore, it is but incidental and wholly derivative. Moreover, you are to hear soon from the man who is best qualified to discuss various phases of ancient science, on the basis of his own pioneering researches; I am not competing with Professor Neugebauer. The sole excuse for the inclusion of the present paper in your group is this: There were certain features in proto-historic Mesopotamia which tended to encourage scientific progress.

* Professor of Semitics, University of Pennsylvania; Director, American School of Oriental Research in Baghdad.

The results happen to constitute the first recorded evidence of scientific performance known to us today. To this extent we are justified in touching here upon the beginnings of science, including the natural sciences. But it should be made clear at the outset that this paper is concerned not so much with the results as with the background; a combination of circumstances conducive to concerted scientific activity rather than the subjects affected by that activity. The background gives us in this instance the essential starting point; it is thus more significant than the immediate achievement.

Our interest, then, will center on a particular cultural stage at which there were at work forces that led to extensive scientific developments; forces which provided the predisposition, so to speak, to these developments. Accordingly, we shall ignore such sporadic achievements of a still more remote age as the invention of the wheel, the introduction of the brick-mold, and perhaps the use of instruments in effecting accurate geometric designs on very early forms of painted pottery. We may have here Mesopotamian inventions which were to play important parts in the eventual progress of engineering, architecture, and perhaps geometry. But these inventions represent isolated contributions of discontinuous cultures which scarcely had any immediate bearing on scientific progress. This paper will confine itself to subjects which had a common origin in a well-defined period and area; which involve from the start habits of observation, classification, and analysis; and which enter then and there upon a continuous course of development.

The region to which our inquiry will take us is Lower Mesopotamia, the land of Ancient Sumer. More specifically, it is an area extending southwest from the environs of Babylon, past Uruk—the biblical Erech—and on along the Euphrates to the metropolis of Ur. The time is the middle of the fourth millennium B.C. This is not just a convenient round figure. It will allow a margin of scarcely more than a century, and in a total of well over five thousand years this is not a disproportionate margin of error. We are in a position to establish the time with such accuracy because it falls within a well-stratified cultural period marked off sharply by distinctive material remains. Soon thereafter there begin to appear inscribed records which tie up before long with concrete regnal years and provide thus a basis for absolute chronology.

We get our first inscribed documents from a level dated to shortly after 3500 B.C., one of a long series of strata recovered from the remains of ancient Uruk. It is among these documents, written on clay, that we find a few which represent the earliest known scientific records. That similar records of still greater antiquity will ever turn up outside Mesopotamia is highly improbable. All available evidence points to the conclusion that the scientific notations with which we are concerned were compiled in close association with the introduction of writing itself. To be sure, this evidence applies only to the script of Mesopotamia. But writing in all the other ancient centers of civilization is demonstrably later. In Egypt it was introduced some centuries after it had been evolved in Mesopotamia, and its first appearance in India was later still. As for the script of China, there is nothing to indicate that it was earlier than the second millennium B.C. It follows, therefore, that the scientific notations on our earliest Mesopotamian tablets constitute not only the first evidence of scientific activity in Sumer, but represent also the oldest recorded effort of this kind known from anywhere in the world. With this significant fact in mind we shall now turn briefly to the records themselves.

What is it that would justify the use of the term "scientific" as applied to a few of the oldest inscribed documents from Mesopotamia? The answer is bound up with the character and purpose of these special texts. Each of them contains lists of related entries. But these lists have nothing in common with the customary inventories of a strictly economic nature. They serve an intellectual rather than a material purpose. And yet, they are to enjoy a continuity and distribution which will set them off sharply from the usual run of business documents whose significance is at once temporary and local. The lists in question are destined to be copied and recopied for many centuries and in more than one city and country. Actual examples of such copies, often modified and expanded, but still in a clear line of descent from the oldest prototypes, have been discovered in Mesopotamian sites of much later age, and even in foreign capitals like Elamite Susa. We have thus before us the beginning of a family of documents of a scholarly character which are notable for their continuity, distribution, and purposeful adherence to an established tradition.[1]

In this recording of accumulating experience and the manifest applicability of such records to the needs of cultural centers separated by political, linguistic, and chronological barriers, we have the essential ingredients of scientific performance. Now what science or sciences did this activity embrace? We shall see presently that the primary purpose of the lists under discussion was to aid in the preservation of the knowledge of writing. Before long, philological studies become an added objective, owing largely to the composite ethnic and linguistic background of early historic Mesopotamia. But natural sciences, too, soon come in for their share of attention.

For regardless of the primary purpose of our lists, they happen to include quite early in their history groupings of birds, fish, domestic animals, plants, and the like. It is worth stressing that these compilations presuppose careful observation and imply organization and analysis of the accumulated data.[2] As an element in the cumulative tradition of the land the lists are subject to steady expansion and improvement. What is more, although these texts were calculated originally to serve purposes unrelated to their subject matter, they lead in course of time to the independent study of the subject matter involved. The fields affected are zoölogy and botany, and later on geology and chemistry. The first recognition of all these subjects as so many separate fields of study may be traced back, therefore, to the earliest inscribed documents from Mesopotamia. Interestingly enough, that recognition was due ultimately to the fact that man had just discovered in writing a way to arrest time and was applying all his ingenuity to the task of keeping this discovery alive.

The subsequent progress of the individual sciences just mentioned has to be traced by specialists. We are concerned at present with the initial impetus alone and the time and circumstances in which that impetus was first received. A few details, however, may be brought out in passing. In the light

[1] These facts are brought out clearly by A. Falkenstein, whose *Archaische Texte aus Uruk* (Berlin, 1936) is the basic work on the earliest documents from Mesopotamia; cf. especially pp. 43 ff.

[2] Careful observation is evidenced also by the accurate drawings of the early pictographs, particularly where exotic animals and specific plants were concerned.

of the foregoing remarks botanists will not be surprised to learn that many of the terms which they use today are found in Mesopotamian sources. These terms include "cassia" (cuneiform *kasû*), "chicory" (*kukru*), "cumin" (*kamûnu*), "crocus" (*kurkânû*), "hyssop" (*zûpu*), "myrrh" (*murru*), "nard" (*lardu*), "saffron" (*azupirânîtu*), and probably many others. The zoölogical compilations which are accessible in cuneiform records contain hundreds of names systematically arranged and presented in two columns, the first giving the Sumerian term and the other its Akkadian equivalent.[3] The scholastic tradition in chemistry[4] results in such texts as the one which has come down to us from the second millennium B.C., wherein a formula for glazing pottery is preserved in the guise of a cryptogram so as to remain hidden from the uninitiated.[5] The importance of the natural sciences for the study of medicine is self-evident; it was not lost on Babylonian and Assyrian medicine.

So much for the indirect benefits derived from the lists under discussion. But the primary objective of these compilations was not allowed to suffer in the meantime. On the contrary, the direct results which were achieved with their aid led to an immensely fruitful advance in another field of intellectual progress.

It was stated above that our lists were intended as a means to preserve the newly attained knowledge of script. By the very nature of its origin in concrete pictographs early writing was an elaborate medium consisting of thousands of items. To each new prospective user it represented a code which could not be deciphered without a proper key. The lists were calculated to supply that key. They were analytical catalogues of signs arranged according to form. Inasmuch as each sign was at first a reflection of something specific in the material world, these catalogues were at the same time systematic groupings of related objects; hence their incidental value to the natural sciences, as we have just seen. The immediate purpose, however, of these arrangements was pedagogical; they

[3] See Benno Landsberger (in coöperation with I. Krumbiegel), *Die Fauna des alten Mesopotamien* (Leipzig, 1934).

[4] On this subject cf. R. Campbell Thompson, *A Dictionary of Assyrian Chemistry and Geology* (Oxford, 1936).

[5] R. Campbell Thompson and C. J. Gadd, in *Iraq*, III (1936), pp. 87 ff.

are our oldest manuals for the discipline of education. As pictographs and ideograms gradually took on abstract phonetic values, the study of the script became linked perforce with the study of language. After the Semitic-speaking Akkadians had joined the Sumerians in building up the civilization of Mesopotamia, linguistic studies rose to exceptional heights against this bilingual background.

The deep-rooted respect for scholarly tradition which comes with a sense of dependence on the contributions of the past, implicit in the developments here outlined, had much to do with the unparalleled achievements of ancient Mesopotamia in the field of linguistics. For it meant that the Akkadians, Babylonians, and Assyrians must fall back upon records in the unrelated tongue of Sumer. The knowledge of that language had to be maintained for cultural purposes long after its speakers had lost all political power, even after they had disappeared from the scene altogether. For the first time in history translators are at work to commit their renderings to writing. This activity called for the production of various auxiliary manuals: syllabaries giving the phonetic value, form, and name of each given sign; vocabularies containing the Sumerian pronunciation, ideogram, and Akkadian equivalent of each word or group of words; lists of synonyms, commentaries on selected ideograms, interlinear transliterations with given Sumerian texts, and the like. Nor was this all. The scientific analysis of Sumerian took the form of grammatical works arranged in paradigms according to the parts of speech and explicit down to such minutiae as the place of the accent. Differences in the dialects of Sumerian were carefully noted. And most of the formidable apparatus was available and in use four thousand years ago! It is to this apparatus that we owe our present knowledge not only of the various dialects of Sumerian and Akkadian, but also of such languages as Elamite, Hittite, Hurrian, and Urartian. As linguistic material these languages may be of interest only to a small group of specialists. But as the media for expressing the thought of a large portion of the ancient world over a period of three millennia—a period one and a half times as long as the whole of the present era—they have a deep significance for the entire civilized world.

The foregoing outline has had as its main theme the demonstration that many forms of scientific progress in Mesopotamia were influenced and linked together by a scholarly tradition which was in turn the by-product of the invention of writing. Our survey has failed, however, thus far to include mathematics and astronomy, two fields for which Mesopotamia has long been celebrated, and is so now more than ever owing to the researches of Professor Neugebauer. It goes without saying that these subjects were affected no less than the other disciplines by the same forces which made for a broad cultural advance in general. But the primary cause of the extraordinary development of mathematical and related studies in Mesopotamia is to be sought, I believe, in conditions which antedate the introduction of writing. In fact, I would add, the origin of writing as well as the interest in mathematics are to be traced back, in this case, to a common source. This source will be found inherent in the society and economy of the prehistoric Sumerians.

We know today that the Sumerians got their idea of writing from the cylinder seals which they engraved with various designs to serve as personal symbols. These symbols came to be employed as marks of identification for religious and economic purposes, for example, with temple offerings. In this representational function the old designs develop into concrete graphs for humans, animals, plants, and so forth, and thence for temples, gods, and cities. The graphs are then associated in each instance with specific words. The gap between picture and word is bridged. Gradually means are devised to express not only complete words but also component syllables, the advance leading thus from the concrete to the abstract. At length writing is perfected to function as a flexible medium for the recording of speech and thought.

When we look back now on the successive interlocking stages in this complicated process, which has been sketched here in its barest outlines, an interesting fact will emerge. The early Sumerians had not set out at all to invent writing. They were carried to this result by a combination of peculiar circumstances. The outcome had scarcely been planned or foreseen. The achievement of the discoverers lay chiefly in their ability to recognize and seize their opportunity. This they did with truly remarkable ingenuity and perseverance. That they had the opportunity to begin with was due, however, to the way in which their society functioned. This system can now be reconstructed from a wealth of diversified evidence. Only a rough summary can be attempted at present.

We have seen that the immediate ancestor of Mesopotamian writing was the cylinder seal, which was first and foremost the Sumerian's mark of ownership. Impressed on clay or cloth, it served to safeguard in the eyes of God and man one's title to possessions or merchandise. We have here a clear indication of a strongly developed sense of private property and thereby of individual rights and individual initiative.[6] The curious shape of the cylinder seal, original with the Sumerians, is explained by its use as a mark of individual ownership. For such cylindrical objects are well suited to cover uneven surfaces with their distinctive design.[7]

Wholly consistent with this economic origin of writing is the fact that the earliest written documents are given over to temple economy. Later texts branch out into the field of private business. Both these uses testify independently to the importance attaching to property rights. Records of a non-economic character are the last to appear, except for the lists discussed above which served as direct aids to writing. The first inscribed documents were used, accordingly, for economic ends, precisely as the cylinder seals themselves. It is easy to understand why the oldest pictographs were so often identical with the designs on the seals.

It follows that Mesopotamian writing, and hence the first script known to man, was the unforeseen outgrowth of a social order which was founded on a recognition of personal rights. This basic feature of Sumerian society is attested overwhelmingly in cuneiform law, perhaps the most characteristic and the most abundant expression of ancient Mesopotamian civilization. In the last analysis this law rests on individual rights. It is not surprising, therefore, that proof of ownership becomes a vital necessity under this system. Incidentally, the rigid requirement of such proof is the main reason for the hundreds of thousands of legal documents recovered from the buried sites of Mesopotamia; the forces responsible for the introduction of writing continued thus as the primary factor in the subsequent

[6] Cf. E. A. Speiser, "The Beginnings of Civilization in Mesopotamia," *Supplement* to the *Journal of the American Oriental Society*, No. 4 (Vol. 59, 1939), pp. 17 ff. (esp. pp. 25-28).

[7] See H. Frankford, *Cylinder Seals* (London, 1939), p. 2.

popularity of script. The law applies to ruler and subjects alike. The king is at first no more than a "great man," as is shown by the Sumerian etymology of the term as well as the form of the corresponding pictograph. He may become the administrator of a vast empire, but even then he is still the servant, not the source of the law, and is responsible to the gods for its enactment. There is here no encouragement of absolute power. Law codes are the constitution which guides the ruler and safeguards the subjects. We have seen that this system is capable of promoting cultural progress on an extensive scale. Its inherent vitality is evidenced by the ease with which this order maintains itself for thousands of years in spite of a succession of political changes under the Sumerians, Akkadians, Gutians, Babylonians, Kassites, and Assyrians. Nor is further expansion hindered by ethnic or linguistic obstacles in its path; for distant and heterogeneous outsiders are attracted not infrequently to the orbit of the Mesopotamian civilization. Among the newcomers we find the Elamites, the Hurrians, and the Hittites, the last-named a people of European ancestry and Indo-European speech. Incidentally, it is to the influence of Mesopotamia upon the Hittites that we owe today our oldest available records of any Indo-European language. The newcomers proceed to copy the laws, use the script, and enjoy the other benefits of the adopted civilization.

Enough has been said to imply that mathematics and time-reckoning were bound to prosper against this social and economic background. An obvious corollary is preoccupation with metrology, with the result that Mesopotamian weights and measures spread eventually beyond the domain of the parent culture. But the technical features of these disciplines do not lie within the scope of the present paper;[8] neither do they fall within the competence of this reader.

To sum up, there existed an intimate relation between scientific progress in Mesopotamia and the source of historic Mesopotamian civilization. Underlying all was a social order resting on the rights of the individual, embodied in a competitive economy, and protected by the supreme authority of the law. This system brought about the evolution of writing, henceforward a decisive factor in the advance of civilization and its

[8] Note the article by V. Gordon Childe, on "The Oriental Background of European Science," *The Modern Quarterly*, I, Number 2 (1938), pp. 105 ff.

diffusion past the changing ethnic and political boundaries. We have here the essentials of a truly cosmopolitan civilization notable for its assimilatory power and a science broad in scope and balanced through the inner unity of its many branches.

Would this story of scientific development have differed appreciably under another type of civilization? The answer is hinted in one of history's most magnificent experiments. The one center possessing a culture of comparable antiquity but dissimilar social and economic background was Egypt. Here the king was a god and as such the absolute ruler and titular owner of all that his realm contained. Under this concept of government there was no room for the recognition of private ownership of property and the all-embracing power of the law. The pharaoh was dictator of a state genuinely and thoroughly totalitarian. The pyramids bear lasting and eloquent testimony to his enormous authority.

We are not concerned here with the respective merits of two contrasting forms of government. Our interest is confined for the present to the effect of coexistent civilizations upon the progress of science in the two centers under comparison. The perspective of more than five thousand years cannot but deepen our appreciation of the debt which modern life owes to both Egypt and Mesopotamia. By the same token, however, we are able now to view objectively some of the differences between their respective achievements.

The established superiority of Mesopotamian mathematics may be attributed, in part at least, to the stimulus of the local economy, so different from the Egyptian. Opposed concepts of property ownership and the fundamental rights of the individual were responsible for the intensive pursuit of legal studies in the one instance and their subsidiary rôle in the other. The astounding accomplishment of Mesopotamia in the field of linguistics had no adequate counterpart in Egypt. We have just seen that in Mesopotamia progress in linguistic studies, not to cite now other branches of science, was linked intimately with the development of writing. But was not Egyptian writing a correspondingly potent factor?

If this question cannot be answered with complete confidence it is largely because the origin of the Egyptian form of script is still open to conjecture. Some details, however, are certain and beyond dispute. The earliest inscribed records of

Egypt are some centuries later than the first written documents of Mesopotamia. In Sumer we can follow the successive paleographic stages step by step, whereas in Egypt the formative period of writing seems to have been very short indeed, to judge from the available material. Moreover, writing left in Sumer a clearly marked trail which leads back to a specific social and economic set-up; in Egypt there is no such demonstrable relationship. Because of all these facts, and in view also of commercial and cultural links known to have connected Egypt and Mesopotamia at the very period under discussion, it is logical to assume that Egypt imported the idea of writing from Mesopotamia. Differences in the form and use of the signs would correspond, then, to the existing differences in the art and languages of the two cultural centers. On present evidence, any other assumption would leave far too much to coincidence.[9] In the final analysis it is not so much a question of the mere use of script as of the conditions responsible for the original emergence of writing.

At all events, Egyptian writing, regardless of its origin, inevitably played its part in the notable progress of Egyptian science. What we miss here, however, is the scope and inner unity of scientific advance which we found to be so characteristic of Mesopotamia. That unity was the product of a tradition which is traceable ultimately to a particular concept of life. In totalitarian Egypt a different set of values attached to life and government and tradition. Is this the reason for an effort that seems more sporadic, greater perhaps in its power of concentration on specific objectives, but also more conspicuous for its omissions? Over a period of millennia this appears to be a justifiable comparative appraisal of the results achieved in the field of science by the two oldest historic civilizations.

[9] Cf. Speiser, *op. cit.*, 22, note 12, and Siegfried Schott, in Kurt Sethe's *Vom Bilde zum Buchstaben* (1939), pp. 81 ff.

UNIVERSITY OF PENNSYLVANIA
BICENTENNIAL CONFERENCE

Some Fundamental Concepts in Ancient Astronomy

By

OTTO E. NEUGEBAUER, PH.D., LL.D.*

1. THE aim of this lecture is not to give any kind of complete survey of the fundamental ideas or methods of ancient astronomy but, on the contrary, to show how one single fact, the variability of the length of the days, influenced the structure of ancient astronomy. I choose this kind of approach because I am convinced that real progress in the study of the history of science requires the highest specialization. In contrast to the usual lamentation, I believe that only the most intimate knowledge of details reveals some traces of the overwhelming richness of the processes of intellectual life.

The variability of the length of the days connects two fundamental groups of problems: the variability during the year leads to the problem of the determination of the orbit of the sun; the variability with respect to the geographical latitude involves the question of the shape of the earth. Both problems are not only very intimately connected, but both require for adequate treatment the creation of a new mathematical discipline—spherical trigonometry. No one of these three groups of problems—ancient theory of the movement of the sun, determination of the shape of the earth, and history of trigonometry—could be adequately discussed in a single lecture. I will therefore confine myself to a short report about some of the questions involved, which are, I believe, in a certain sense typical for the situation faced by the ancient mathematician, and I will discuss only those methods which are of essentially *linear* character. This means that I shall disregard the mathematical part of the problem, the history of spherical trigonometry,[1]

*Professor of Mathematics, Brown University.

[1] The branch of the development where *trigonometric* methods are involved will be discussed in a forthcoming paper by Olaf Schmidt, Brown University. In the following the treatment of these problems by stereographic projection as

and, instead, emphasize an earlier stage of our problem, whose importance for different problems in ancient astronomy has not been fully acknowledged.[2]

2. When we talk about the "length of the days" we must briefly discuss concepts and methods of measuring time. We all have some feeling of homogeneous time as a kind of equidistant scale, well adapted to measure the events in the observed world. I will not discuss the fact that this a priori concept of homogeneous time is doubtless due to the fortunate fact that we are living on a celestial body which moves under almost the simplest possible conditions (the so-called two-body problem) and that celestial mechanics shows that only with very little change in the original distribution of masses and velocities our aspect of the sky would be about the same as the aspect of the lights of a large city from a roller-coaster,[3] where nobody would create such nice concepts as our day and night and their smaller parts. But even under the ideal conditions given on our planet, it took more than two of the four millennia of known history to develop such a simple concept as an "hour" of constant length.

It is well known that "hour" meant in ancient and medieval times one-twelfth of the actual daylight from sunrise to sunset, so that "one hour" in January and August, and in Alexandria and Rome, had very different lengths. From our point of view the first question may be: How is it possible to arrive at such an obviously inconvenient definition of time? However,

---

given by Ptolemy in his "Planisphaerium" (opera, vol. II, pp. 225-259) is completely disregarded.

[2] It must be emphasized that Honigmann in his book *Die sieben Klimata und die πόλεις ἐπίσημοι* (Heidelberg, Winter, 1929) recognized for the first time the relationship between the problem of the "rising-times" treated here and ancient geography. Independently Olaf Schmidt discovered the importance of these questions for the ancient geometry of the sphere, especially in Theodosius. These two sources, together with my own investigations on Hypsicles, directed my attention to the "linear methods" in Greek and Babylonian astronomy and their relationship.

[3] Cf. e.g., the results of Hill (*Coll. Works*, I, p. 334 f.) and Poincaré (*Méth. Nouv. Méc. Cél.*, I, p. 109) which show that only a slightly different initial situation would cause our moon to move in a curve of oval shape in the main part, but with a loop at each end of the longer axis, such that the moon would appear half six times during one revolution. In the neighborhood of full- and new-moon, the moon's velocity would be about the same as now, but around the loops the movement would be almost zero. I wonder what kind of time concept would be proved to be a priori by the philosophers of the dwellers on such a moon.

formulating the question in this way prevents access to the solution. We must not ask who invented the hours of unequal length (the so-called "seasonal hours"), but we must find the causes which finally enforced the creation of such a highly artificial concept as an hour of constant length ("equinoctial hour"). Actually no simple observable phenomenon exists which may give a time-scale with equidistant intervals: vessels of a very special shape only give constant quanta of water-outflow, the shadow changes according to complicated trigonometrical functions, the length of the day changes in rates which are far from being linear, and the stars shift from night to night, and there do not exist ancient clocks exact enough to show the regularity of their movement. However, all those irregularities were just small enough to make *linear* approximations not entirely impossible. The brief discussion of their character and relationship is the topic of this lecture.

3. Apparently the most natural division of day and night is the division into two, three, or four parts. The bisection gives noon and midnight, the thirds are the "watches" in Babylonia,[4] the quarters the "watches" in Egypt.[5] The variation in the lengths of the nights in those southern latitudes is so small that no one needed to worry about the constancy of these watches. How uninterested the Egyptians were in the change of the astronomical seasons is emphasized by the fact that they subdivided their year not in four but in three agricultural seasons: the period of inundation, the reappearance of the fields from the inundation, and the harvest.[6] Obviously primarily agricultural societies do not need any kind of precise definition of homogeneous time; and even in periods where a finer subdivision is required, the older custom of treating day and night separately has been kept in use. Therefore the Egyptian "hours," which can be shown to exist since about 2000 B.C.,[7] are typically seasonal hours of one-twelfth of the day and one-twelfth of the night each. In Greek literature these "hours" do not appear earlier than in Hellenistic times.[8]

[4] Cf. e.g., B. Meissner, *Babylonien und Assyrien*, II, p. 394 (Heidelberg, Winter, 1925).

[5] K. Sethe, *Die Zeitrechnung der alten Aegypter* (Nachr. Ges. d. Wiss. Göttingen Phil-hist. Kl. 1919, p. 287 ff. u. 1920, p. 97 ff., p. 127).

[6] Sethe, *Zeitrechnung*, p. 294.

[7] Sethe, *Zeitrechnung*, p. 111.

[8] The oldest occurrence of "hours" as a well-determined time measure seems

A very different but also very primitive method of counting time has been developed in Mesopotamia. We know that as early as in Sumerian times[9] there existed a distance-unit named *danna*, which may be translated as "mile," corresponding to about seven of our miles. This unit was used for measuring longer distances and became in this way quite naturally also a time-interval: the traveling time for such a distance. If we suppose this slight change in the meaning of the word "mile," it is immediately intelligible how a day or a night could be expressed in "miles." But the origin of these "time-miles" from measuring distance has never been forgotten, and therefore time measurement in "miles" became a homogeneous one, independent of the changing length of the day during the seasons. When later, I may say some time in the first part of the first millennium B.C.,[10] Babylonian astronomy made its first steps to a more systematic recording of celestial phenomena, this length-measure "mile" was transferred to celestial distances too, in the simple way that the number of miles contained in one day was made equivalent to *one* revolution of the sky. Because one day contained twelve of these itinerary miles, the circumference of the sky also became twelve miles. And because the mile (*danna*) has been subdivided in thirty UŠ (the meaning of UŠ is very significant, simply "length"), the length of the main circle of the sky was divided into $12 \cdot 30 = 360$ parts. This is the origin of our "degrees" and the custom of modern astronomy of measuring time in degrees.[11]

to be in the writings of Pytheas (time of Alexander the Great), quoted by Geminus VI, 9 (ed. Manilius p. 70, 23 ff.); hours are frequently used by Geminus (ca 100 B.C.), Vitruvius and Manilius (time of Augustus). For further literature, see Kubitschek, *Grundriss der antiken Zeitrechnung* (Handb. d. Altertumswiss. I, 7) p. 179. Herodotus (400 B.C.) is often quoted for mentioning the Babylonian "hours" (II, 109) but this sentence has been considered to be an interpolation [recently by J. E. Powell, Classical Review 54, 1940, p. 69 (without knowing an older attempt in the same direction, mentioned by Kubitschek p. 178, note 1)] but, I think, without sufficient reason.

[9] Oldest example from Tello, period of Agade (about 2400 B.C.) published by Fr. Thureau-Dangin, *Inventaire des Tablettes de Tello conservées au Musée Impér. Ottoman*, Paris, 1910 ff., 11, 1175.

[10] Cf. A. Schott, *Das Werden der babylonischen Positionsastronomie*, Zeitschr. d. Deutschen Morgenländischen Gesellschaft **88** (1934), 302 ff. and his review of Gundel, *Hermes Trismegistos*, in Quellen u. Studien z. Geschichte d. Math., Abt. B., vol. 4 (1937), p. 167 ff.

[11] Cf. O. Neugebauer, *Untersuchungen zur Geschichte der antiken Astronomie III*, Quellen u. Studien z. Geschichte d. Math., Abt. B., vol. 4 (1938), 193 ff.

In modern literature those "miles" are very misleadingly named "double hours" because they correspond actually to two of our time units.[12] But the ancients were well aware of their origin; e.g., Manilius (time of Augustus) speaks correctly about *stadia*, i.e., miles, in his famous astronomical poem.[13] These Babylonian time-distances appear frequently in Greek astronomy and give clear evidence of the important influence which Bablyonian astronomy exercised in the ancient world.[14]

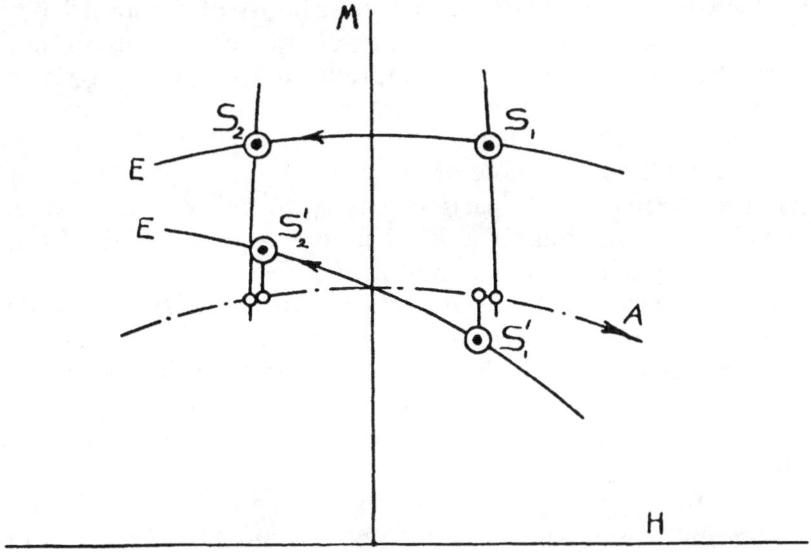

FIG. 1.

4. It may seem that with this (certainly unconscious) creation of a homogeneous time all trouble was over, but the real difficulties begin with the introduction of the concept of homogeneous time: we have to express the natural time intervals, as day and night, by the lengths of some constant time intervals. This problem has two different aspects: first, the practical one

12 G. Bilfinger, *Die babylonische Doppelstunde*, Stuttgart, Wildt, 1888.

13 Manilius, *Astronomica III*, 275 ff. (ed. Breiter, p. 74 and p. 88; ed. Housman p. 24 and p. XIII ff.).

14 They appear e. g., in Herodot II, 109 (ca 450 B.C.), in the "Eudoxos-papyrus" (about 3rd cent. B.C.; cf. Pauly-Wissowa, *Real-Enzyklopädie* 6 col. 949), in the Pap. Michigan 151 (*Michigan Papyri*, vol. III, p. 118 ff., a text which I intend to discuss in a forthcoming paper) and implicitly, of course, in the countless places where degrees are used to express time (first instance in Greek is in Hypsicles, about 200 B.C.).

of constructing clocks showing real constant time intervals; secondly, the theoretical problem of finding the rule by which the length of the days, expressed in this constant time interval, changes. In the following we shall be mainly concerned with this second question.

In order to understand fully the problems involved, it may be remembered how "one hour" is defined today. The simple definition, "One hour is the twenty-fourth part of the time from noon to noon," i.e., from one meridian-passage of the sun to the next, is not sufficient to obtain hours of constant length for two reasons. First, the velocity of the sun is not constant. Secondly, even under the assumption that the sun travels the same part of its orbit every day, the fact that this orbit (the "ecliptic" E in fig. 1) has an inclination of more than twenty-three degrees to the plane of our daily rotation (the "equator" A) implies that equal parts on the ecliptic do not cross the meridian (M) in equal lengths of time.[15] In order to avoid both these irregularities (which combined give a rather complicated effect) modern astronomy introduced an artificial body called "mean sun" which moves with the constant *average* velocity of the "true sun" and which has the *equator* as orbit, and not the ecliptic. The twenty-fourth part of these *artificial* days lasting from meridian passage to meridian passage of the *mean* sun is our familiar "hour."[16]

5. Both sources of this complication in the definition of "time" were well known to ancient astronomy. The direct observation of the variability in the sun's daily path is of course far beyond the capacity of any kind of instrument available to the ancients. However, they realized that the number of days elapsing between the vernal equinox and the summer solstice is not the same as the number of days between summer solstice and autumn equinox, between autumn equinox and winter solstice, and winter solstice and vernal equinox, and that these four points divide the year into unequal parts. The method of taking this observation into consideration is very characteristic of the ancient astronomical systems.

[15] As an example, in fig. 1, are shown two equal parts S_1S_2 and $S'_1S'_2$ of the ecliptic which the sun may travel in a day, one at summer solstice (E practically parallel to A) and the other at the vernal point (E maximal inclination to A).

[16] The difference between true and mean solar time (the so-called "time equation") reaches a maximum of about ±15 minutes.

The oldest Bablyonian system, which must have been created earlier than about 200 B.C., introduced an artificial sun, moving in two parts of the year with a different velocity, suddenly jumping at two well-defined points of its path from one velocity to the other (fig. 2a). These velocities and jumping-points were

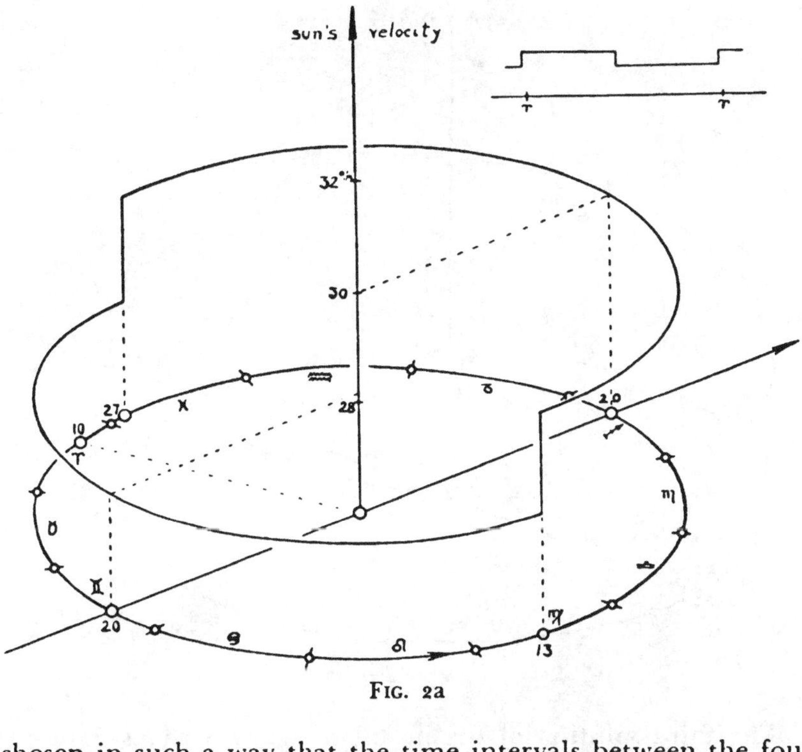

FIG. 2a

chosen in such a way that the time intervals between the four points in the year, mentioned above, are just the times required by observations. I think that this purely mathematical construction shows the surprisingly high level of this late-Babylonian astronomy.

Probably very soon[17] after this first attempt to describe mathematically the movement of the sun, a second theory was developed in Babylonia, where the required change in the sun's velocity was represented by an apparently more natural con-

[17] An attempt of P. Schnabel to date the origin of the two Babylonian systems exactly (*Berossos und die babylonisch-hellenistische Litteratur*, Leipzig, Teubner, 1923, p. 223 ff.) is often quoted in the literature, but it is based on assumptions which can easily be proved to be wrong.

struction, namely by the assumption of *linearly* changing velocity (fig. 2b). The reason why this model is later than the first mentioned is a purely mathematical one, because the further consequences of this second assumption become much more complicated than the first case.[18]

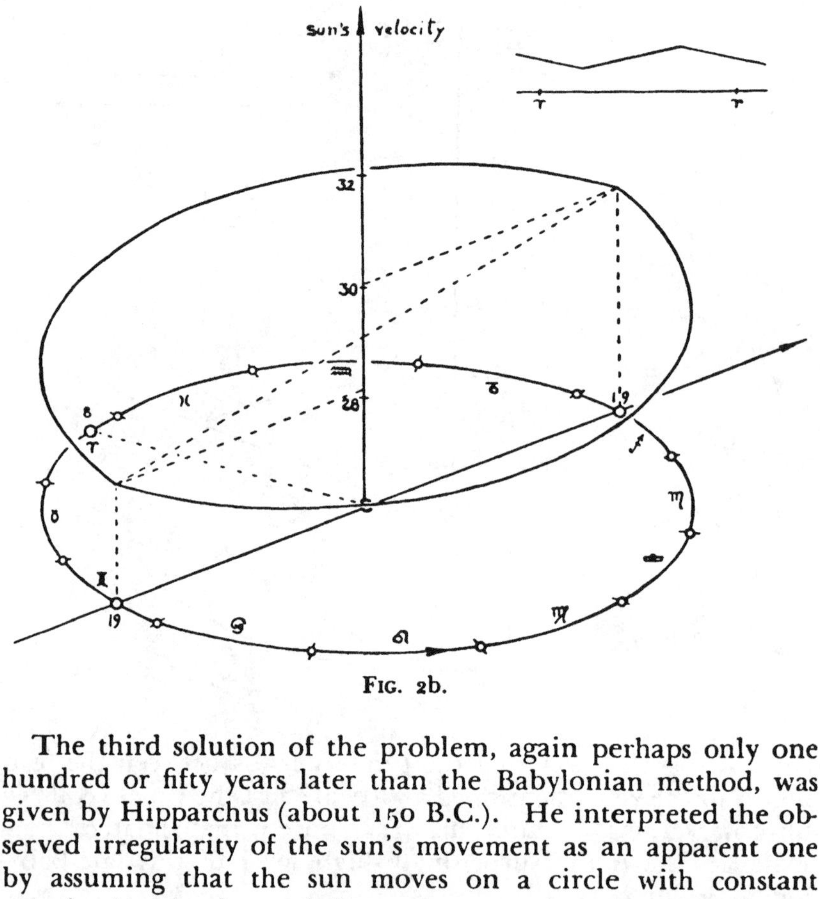

FIG. 2b.

The third solution of the problem, again perhaps only one hundred or fifty years later than the Babylonian method, was given by Hipparchus (about 150 B.C.). He interpreted the observed irregularity of the sun's movement as an apparent one by assuming that the sun moves on a circle with constant velocity but is observed from an *eccentric* point. This is the type of astronomical theory which determined the astronomy of the following 1,500 years, in a certain sense doubtless a regression from a pure mathematical method to assumptions about the physical nature of our planetary system (fig. 2c).

6. Let us now consider the second part of the questions in-

[18] Expressed in modern terms: The summation processes which are required in the theory of syzygies become one degree higher in the second theory.

volved in the determination of the length of the days, namely the inclination of the ecliptic. According to ancient custom the "day" began with sunrise or sunset; the second definition was adopted in Babylonia obviously because every new month began with the first visibility of the new moon, which comes just after sunset. The ancient problem is therefore the deter-

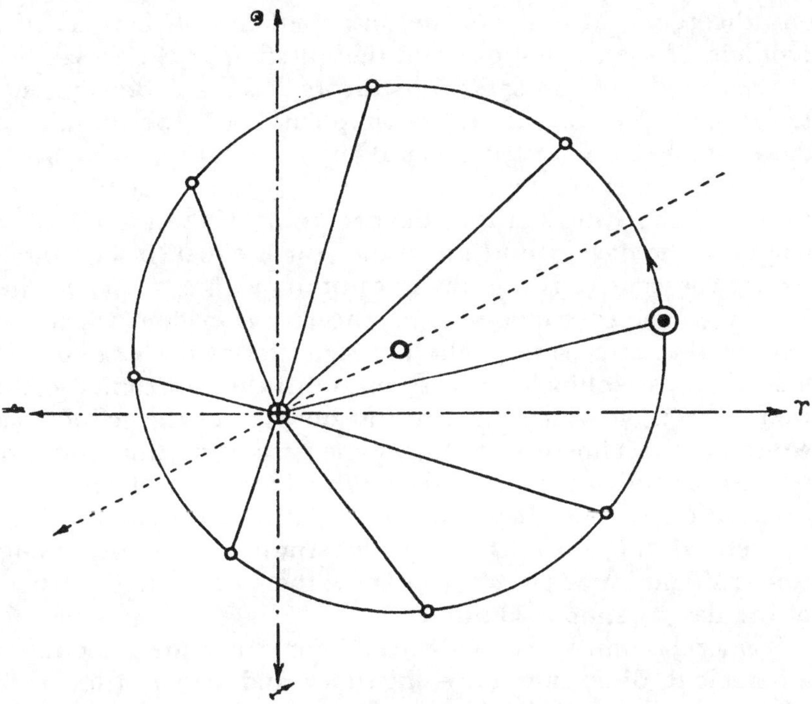

FIG. 2C

mination of the time elapsed between two consecutive crossings of the horizon by the sun. Here the same difficulty occurs as in the above-mentioned case of the crossing of the meridian line by the real sun, moving on the ecliptic and not on the equator; whereas (for a given place) the inclination of the equator to the horizon is constant, the sun's orbit cuts the horizon at continuously shifting angles. The problem is the famous problem of the determination of the "rising-times" (*ἀναφοραί*) in Greek astronomy: to calculate the equatorial arcs which cross the horizon in the same time as a given arc of the ecliptic.

This problem is obviously a problem of spherical trigonometry. Its complete solution can be found in Ptolemy's *Almagest* (ca 150 A.D.)[19] and has profoundly influenced the earlier treatises on the geometry of the sphere from Autolycos (little before 300 B.C.), Euclid (ca 300 B.C.), and others to Theodosius (ca 100 B.C.) and Menelaos (ca 100 A.D.).[20] According to the limitations of this lecture I shall not discuss the history of this part of the theory. However, I must mention one theorem which shows the direct connection between the problem of the rising-times and the question of the variability of the lengths of the days. This theorem is the following one. Let us consider, for the sake of simplicity, only the twelfths of the ecliptic, the so-called zodiacal signs. Let $\alpha_1, \alpha_2, \ldots, \alpha_{12}$ be the rising times of the first, second, . . . , twelfth sign, respectively. Then, if the sun is at the beginning of the i-th sign, the length of the day at this time of the year is equal to the sum of the six consecutive rising times beginning with α_i, i.e., $\alpha_i + \alpha_{i+1} + \ldots + \alpha_{i+5}$. The correctness of this theorem is evident when you remark that after sunrise the i-th sign crosses the east-horizon first, then the following and so on, until the sun comes to the west-horizon, at which moment just one half of the ecliptic has crossed the east-horizon.[20a] During the time from sunrise to sunset just six zodiacal signs cross the east horizon, and this is the proposition of our theorem. This relation explains the high interest of the ancients in the determination of the rising-times: if you know the α's, you know the corresponding length of the day by simple addition.

This relationship is fundamental for the understanding of all ancient discussion of rising-times and day-lengths. It is of course treated in Ptolemy's *Almagest*, where a table of the rising-times for ten different latitudes is given,[21] with intervals of ten degrees. As I mentioned before, this table is calculated by using spherical trigonometry and therefore represents correctly the rather complicated relation between the sun's position in the zodiac and the corresponding rising-times of the ten-degree arc. This relationship is shown in the case of the

[19] Ptolemy, Almagest II, 7 and 8. An excellent treatment of the problem corresponding to our time-equation is given by A. Rome in *Ann. Soc. Sci. de Bruxelles*, sér. 1, 59 (1939), 211 ff.

[20] Cf. notes 1 and 2.

[20a] Here, as everywhere else in this paper, the difference between sidereal time and solar time is neglected for the sake of simplicity.

[21] *Almagest*, II, 8.

latitude of Alexandria in fig. 3. The characteristic property of this curve is the secondary minimum between the two maxima, a kind of asymmetry which gets worse with increasing geographical latitude.

This exact shape of the curve was obtained only by using trigonometry. However, we know very interesting older attempts to describe the rising-times as functions of the sun's

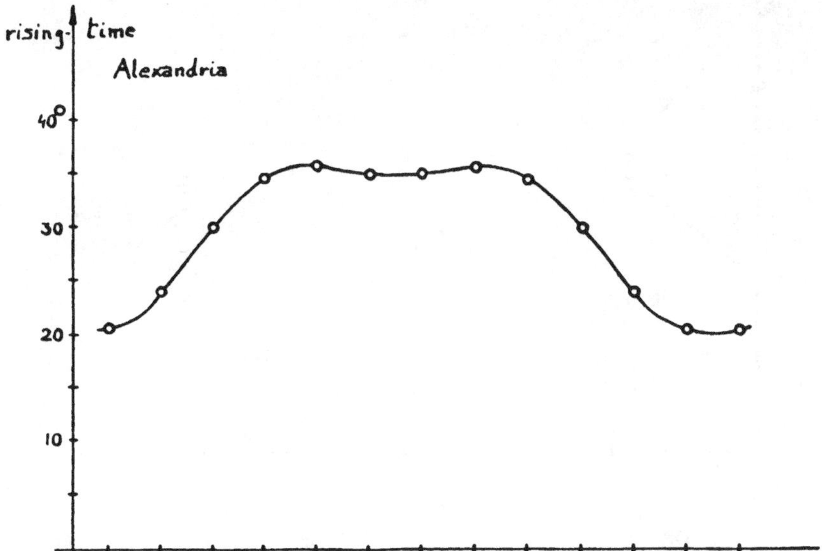

FIG. 3

positions. There exist two different types, one (A) represented by Hypsicles[22] (ca 200 B.C.) for the latitude of Alexandria, the second one (B) by Cleomedes (time of the Roman empire) for the latitude of the Hellespont.[23] Both curves are *linear* approximations of the true curve, with the exception that B inserts in the middle of the slanting lines twice the difference (fig. 4). The corresponding theory about the influence of geographical latitude is given for system A by Vettius Valens (about 150 A.D.),[24] for system B by Pap. Michigan 149,[25] who both

[22] New edition in preparation by M. Krause, V. De Falco, and myself.

[23] Concerning this location see my paper "Cleomedes and the meridian of Lysimachia," accepted for publication in the *Am. J. of Philology*.

[24] I, 7 (ed. Kroll, p. 23).

[25] *Michigan Papyri*, vol. III, p. 63 ff. espec. p. 103 and p. 301 ff. (about 2nd cent. A.D.).

agree on the general method of defining seven "climata" by the assumption that the rising-times increase *linearly* from clima to clima, Vettius Valens starting from Alexandria, the Michigan papyrus with Babylon as main clima.[26]

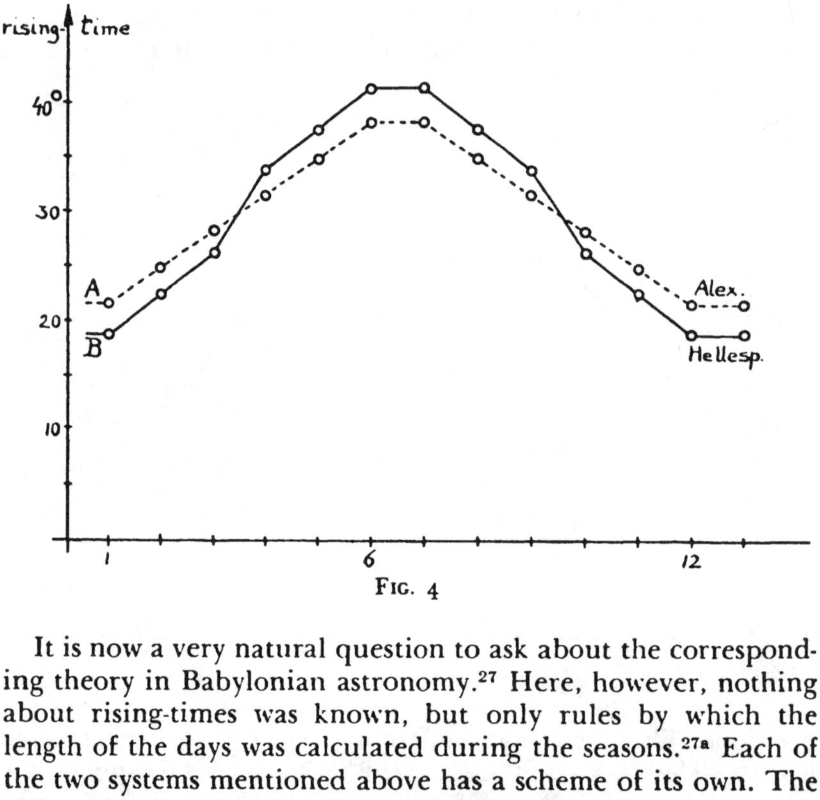

FIG. 4

It is now a very natural question to ask about the corresponding theory in Babylonian astronomy.[27] Here, however, nothing about rising-times was known, but only rules by which the length of the days was calculated during the seasons.[27a] Each of the two systems mentioned above has a scheme of its own. The older one gives (expressed here in degrees) as lengths the following list A, the younger one B:[28]

[26] This follows from a slightly different interpretation of the text, as given by the editors, which requires a much smaller emendation and will be discussed in a forthcoming paper.

[27] Honigmann has already tried to restore the Babylonian rising-times and has discovered that Vettius Valens and Manilius refer to this latitude (*Mich. Pap.*, III, p. 313). He was apparently much disturbed by not quite correct information of Schnabel (p. 314) and a wrong hypothesis of Kugler about a Babylonian scheme of day-lengths (p. 317).

[27a] These rules were discovered by Kugler, *Babylonische Mondrechnung*, Freiburg, 1900, p. 77, p. 99.

[28] Unfortunately Kugler reversed the order of the two systems by calling the older one II, the younger I.

A				B			
10° of	length of the day		10° of	8° of	length of the day		8° of
♈	180	180	♎	♈	180	180	♎
♉	200	160	♏	♉	198	162	♏
♊	212	148	♐	♊	210	150	♐
♋	216	144	♑	♋	216	144	♑
♌	212	148	♒	♌	210	150	♒
♍	200	160	♓	♍	198	162	♓

We must only remember the fundamental relation between the lengths of the day and rising-times in order to find a system of numbers with constant difference from which, by the addition of six of them, the day lengths A and B, respectively, can be derived, namely

A:		B:	
	$\alpha_1 = \alpha_{12} = 20$		$\alpha_1 = \alpha_{12} = 21$
	$\alpha_2 = \alpha_{11} = 24$		$\alpha_2 = \alpha_{11} = 24$
	$\alpha_3 = \alpha_{10} = 28$		$\alpha_3 = \alpha_{10} = 27$
	$\alpha_4 = \alpha_9 = 32$		$\alpha_4 = \alpha_9 = 33$
	$\alpha_5 = \alpha_8 = 36$		$\alpha_5 = \alpha_8 = 36$
	$\alpha_6 = \alpha_7 = 40$		$\alpha_6 = \alpha_7 = 39$

Both lists are *linear*, except the double difference in the middle of B, or, in other words, exactly the same, which we knew from the Greek rising-times, mentioned before. The Babylonian list A of rising-times appears explicitly in Vettius Valens[29] and in Manilius.[30]

7. We can summarize our discussion in the statement that the Greek theory of rising-times and variability of day and night is identical with the Babylonian scheme as far as the latitude of Babylon is concerned, and that the Greeks modified these rules in the simplest possible way, namely, linearly, in order to adapt them to geographical latitudes different from Babylon.

It should be mentioned that these *linear* approximations of the complicated actual curve shown in fig. 3 (p. 23) give very satisfactory results for the lengths of the days, at least as far as this can be controlled by the very inaccurate ancient clocks. The proportion 3:2 between the longest and shortest day, adopted in both Babylonian systems, agrees very well with the actual duration of light at Babylon in the summertime,[31] but this custom of characterizing the latitude of Babylon by the

[29] I, 7 and 14 (ed. Kroll, p. 23 and 28).

[30] Cf. note 13.

[31] Kugler, *Sternkunde u. Sterndienst in Babel*, I, p. 174, II, p. 588; Schaumberger in *Ergänzungen*, p. 377.

proportion 3:2 is the reason for a strange deformation of the ancient world-map, namely, that of placing Babylon at 35° n.l. (instead of about 33½), a misplacement which affected the map of the eastern part of the oikumene very much. For this latitude of 35° is the immediate result of the theory given by Ptolemy, based on the proportion 3:2, which is trigonometrically correct, but neglects all atmospheric influences in the duration of the light-day, which are unconsciously included in the Babylonian values.

8. The theory of the rising-times has one more very important application in ancient astronomy, as far as I know entirely overlooked by modern scholars. This is the question of determination of the length of invisibility of the moon around new-moon. This question is of highest importance for the oriental civilizations in which the calendar was regulated by the actual reappearance of the moon in the evening one or two days after astronomical new-moon. In order to understand these relations between rising-times (or here better, setting-times) and the visibility of the moon, we need only remark that this visibility not only depends on the distance between sun and moon in the ecliptic but also on the inclination of the ecliptic with respect to the horizon. If the ecliptic crosses the horizon almost vertically, obviously a much smaller distance between sun and moon is required in order to make the moon's crescent visible in the dusk than if the ecliptic lies more horizontally and the sun and moon set almost simultaneously.

We know from investigations by Kugler, Weidner,[32] and others that Babylonian astronomers were concerned with the problem of the dependence of the invisibility of the moon on the seasons at a very early date,[33] when even the variability of the length of the days was assumed to be linear. Correspondingly, the first attempt to estimate the time between setting of sun and moon was very unsatisfactory too, namely the assumption of simple proportionality with the duration of the night.

We know very little about the further development of this question in Babylonia, but I think that a chapter in Vettius Valens may give some information. At any place where we are able to check his reports he seems to be very well informed

[32] Kugler, *Sternkunde u. Sterndienst in Babel, Ergänzungen*, p. 88 ff.; E. F. Weidner, *Alter u. Bedeutung d. babyl. Astron.*, Leipzig 1914, p. 82 ff.

[33] Schott l.c. note 10, p. 310.

about Babylonian sources, which he quotes explicitly.[34] I think, therefore, we may assume his chapter I,14 as essentially Babylonian; here he states that for the latitude of Babylon[35] the elongation of the moon from the sun at the moment when the moon becomes invisible is one-half of the rising-time of the corresponding zodiacal sign.[36] This rule is still a strong simplification of the actual facts, but reveals on the other hand the full understanding of the fact that the problem of the moon's invisibility around new-moon requires the consideration of the change of the ecliptic position with respect to the horizon.[37]

The last step in the development of this theory before the complete solution by spherical trigonometry can also be found in cuniform texts, but only in the most elaborate system. Here we find an almost perfect solution of the problem, perfect at least as far as observations with very inaccurate instruments are able to control. Here, first of all, the inequalities in the movement of sun and moon are taken into consideration, furthermore the deviation of the moon from the ecliptic (its "latitude"), including an estimate of the influence of the twilight. Finally the rising-times are used in order to transform the ecliptic coördinates of the moon into equatorial coördinates or into "time."[38] We have here a very impressive example of how, by an ingenious combination of linear approximations and their iteration, a very accurate solution of a problem which seems to belong entirely to the realm of spherical trigonometry

[34] IX, 11 (ed. Kroll, p. 354).

[35] This is proved by the fact that the values he gives as examples are exactly the Babylonian values for the rising-times.

[36] I, 14 (ed. Kroll, p. 28). Details will be discussed in a forthcoming paper.

[37] Vettius Valens discusses in chapter I, 13 (ed. Kroll p. 28) the closely related problem of the daily retardation of the moon's rising and setting with respect to sunrise and sunset. The method is purely linear and based on very rough approximations, but mentioned earlier in Pliny H. N. (first cent. A.D.) and later in the *Geoponica* (6th cent. A.D.). These texts have recently been discussed by A. Rome in vol. II, p. 176 of the work of Bidez and Cumont, *Les Mages hellénisés* (Paris 1938), because the Geoponica refers the method to Zoroaster. In the light of the discussion in the work of Bidez and Cumont, a Babylonian origin would be very possible. The method of expressing fractions, however, is the Egyptian one, which speaks strongly for Egyptian origin in spite of the fact that it is more difficult to understand how Egyptian methods could be connected with the doctrines of the mages.

[38] This fact was first discovered by Schaumberger (*Ergänzungen zu Kugler*, cf. note 31, p. 389 ff.) but only by using modern calculations. The relation to the Babylonian rising-times will be discussed in a forthcoming paper.

can be obtained. However, only careful historical investigation of many scattered facts shows that the high building of ancient spherical astronomy and geography is erected on the ground of age-long older attempts and experiences.

9. Our sources are not sufficient, or at least not sufficiently well investigated, to answer the question about the historical origin of the problem of the rising times of ecliptic arcs. It is possible that independent attempts have been made to determine the variability of the lengths of the days directly. One interesting suggestion has been made by Pogo in his investigations on Egyptian water clocks.[39] We have examples of such clocks since the eighteenth dynasty (ca 1500 B.C.) containing inside different scales in order to subdivide the day in twelve parts at the different seasons of the year. Pogo could explain the arrangement of the scales by the following assumptions: let Δ denote the difference between the longest and shortest day, then the increase of the length of a day in the first month after the winter solstice has been assumed to be $\Delta/12$, in the second $2\ \Delta/12$, in the third and fourth $3\ \Delta/12$; $2\ \Delta/12$ in the fifth, $\Delta/12$ in the sixth and correspondingly in the decrease.[40] It can easily be shown that this rule is equivalent to the newer Babylonian scheme, mentioned above as B, whereas the older one (A) would correspond to the coefficients $\Delta/18$, $3\ \Delta/18$, $5\ \Delta/18$, respectively.

Pogo's remarks would speak in favor of the assumption that the attempt to characterize directly the rule of the variability of the length of the day was the first step in our group of problems—an assumption which sounds in itself natural enough. But it must not be forgotten, on the other hand, that in the Babylonian astronomy the connection between the length of the days and the visibility of the moon, which involves the rising-times, was established very early, as we have seen above. And finally, one large group of questions has been neglected almost entirely, namely, the methods of determining time and geographical latitude by sun dials.[41] It seems to me

[39] A. Pogo, "Egyptian water clocks," *Isis* 25 (1936), p. 403 ff.

[40] *Isis* 25, p. 407 ff.

[41] This method is well known from Egyptian, Greek, and Roman sources. Their existence in Babylon has been proved by E. F. Weidner (*Am. J. of Semitic Languages and Lit.*, vol. 40 (1924), p. 198 ff.). I did not realize until recently that texts which I published as "generalized reciprocal tables" are actually "gnomon texts" of a type a little more developed than Weidner's texts

therefore better not to propose any definite solution of the earlier history of our problem but to emphasize the fact that here is a large field open for and deserving our investigation, if we are interested in understanding the creation of our time scale.

(*Mathematische Keilschrift-Texte,* Quellen u. Studien z. Gesch. d. Math. Abt. A vol. 3, I, p. 30 ff.).

UNIVERSITY OF PENNSYLVANIA
BICENTENNIAL CONFERENCE

Medicine and Surgery in Ancient Egypt

By

HERMANN RANKE, PH.D.*

OF ALL Egyptian medical literature—which must have been very extensive indeed—we have only five long and well-preserved texts, and all of these were written within a period of about five hundred years. Thus we are entirely unable to even sketch a history of the development of Egyptian medicine.

What I can give you within the thirty minutes allotted to me is only a brief outline of what seem to me the most essential features of this literature.

I have to pass over the references in tomb inscriptions of the Old Kingdom, from which we learn that medical texts, written on papyrus, were in use during the middle of the third millennium B.C., and that, at the same time, the medical profession was classified as to rank and also as to specialized fields (such as "court eye physician," "body physician," "dentist," and others).

I have to pass over also the earliest preserved fragments of medical texts of the Middle Kingdom—that is, of the early part of the second millennium—one of which deals with women's diseases, while another contains the only known Egyptian text of a veterinary nature and deals with different diseases of bulls.

Even of those five well-preserved texts which I mentioned, I shall choose only the two largest ones for a closer examination; the so-called Papyrus Ebers, now at the University Library at Leipzig, Germany, and the Papyrus Edwin Smith, now in the possession of the Historical Society of New York and on loan in the Egyptian collection of the Brooklyn Museum. Both these papyri were written in the period following the Middle Kingdom, when the so-called Hyksos ruled over Egypt, i.e., in the seventeenth century B.C.

I have chosen them not only because they are the largest and best preserved of all the Egyptian medical papyri we know,

*Visiting Professor of Egyptology, University of Pennsylvania.

but because they represent the two main groups of Egyptian medical literature. The Ebers Papyrus is chiefly a collection of recipes, intended for the use of the physician. The Edwin Smith Papyrus contains a collection of "cases," intended for the use of the surgeon.

The Ebers Papyrus, a complete roll of over twenty meters in length, was bought in 1873 and edited in 1875, but until recently no reliable translation of it had been published. This gap has now been filled by the masterly English translation of the Norwegian medico-historian B. Ebbell, which appeared in 1937, and which has enhanced our understanding of this important text considerably.

The Ebers contains a collection of no less than 877 recipes which are recommended for a great number of diseases and ailments. This large collection itself is a compilation of a great number of smaller ones which formerly may have had an isolated existence. There are more than thirty such sub-collections, each of which is preceded by the heading "The beginning of . . ." The first group, which is one of the largest, has the rather vague heading, "The beginning of a compilation of remedies," but it deals exclusively with diseases of the belly. Other, very brief ones, contain, e.g., remedies to treat the liver (as Dr. Ebbell shows, they probably refer to jaundice), or "remedies for a finger or a toe that is ill," or "remedies for an ear that does not hear well"—while another very long one contains a "compilation on the eyes," eye diseases playing as great a rôle in antiquity as they do in modern Egypt. The smaller groups have been arranged according to similarity of content, so that five main groups of recipes following one another can be distinguished: internal medical diseases (these take up more than half of the whole), diseases of the eye, of the skin, of the extremities, and diseases of women. Between the last two a group of miscellanea is inserted, partly containing recipes against evils which, according to our feeling, do not require medical treatment. About these I shall say a few words later.

Only in twenty-five cases out of the 877 is a longer diagnosis given, against which a certain recipe is prescribed. Otherwise, the illness or disease is only named—the physician using the papyrus is supposed to know them well—and then the recipe or recipes are given often a number of different ones against the same illness, to choose from ad libitum.

The physician evidently was, at the same time, what we should call the druggist, and had to prepare the remedies which he used for healing. The main rôle is played by vegetable drugs, of which the majority—partly by the efforts of Dr. Ebbell—are known to us, while some still remain unidentified. I can give only a few examples. Of fruits, we find dates, grapes, figs, raisins, sycamore fruit, fruit of juniperus, watermelons, etc. Of vegetables: cucumbers, beans, onions, celery, castor oil, coriander, cinnamon. Of cereals: powder of wheat or barley. Besides, such things as flax seed, leaves of lotus, of cucumber, pine tar, wax, dregs of wine. Of mineral drugs, different salts are used, yellow and red ochre, natron, malachite. Of animal drugs, we have especially milk and the fat of various animals or birds, among others of cats, ibexes, serpents—and also of crocodiles, hippopotami, and even of lions!

The recipes themselves call for pills, potions, onitments, and bandages. The potions consist of water, milk, beer, wine, date wine, juice of acacia or pistacia. We find often the prescription to let them stand in the dew during the night before they are taken, or to serve them "at an agreeable warmth," or "in the morning" or "before going to bed." With the ointments, a rubbing for four days is often recommended, more rarely for ten days, sometimes a rubbing "very early in the morning." Once it is said that the body is to be rubbed and then "placed in the sun."

As you see, these prescriptions are of a very sober and matter-of-fact character, and doubtless served their purpose well. Of some other good, practical prescriptions, I may mention the rinsing of the mouth against an illness of the tongue; the inhaling, through a reed, against coughs; the use of a reed as a drainage tube applied to a bubo or the use of injections against gonorrhea. Suppositories—once with the reassuring statement that they are "cooling"—are recommended against "burning in the anus" and consist either of salt, watermelon, honey, plus a fourth unidentified ingredient, or of fruit of juniperus, frankincense, yellow ochre, cuttle bone, cumin, honey, myrrh, and cinnamon, again plus an unknown x. From the number of drugs you would suppose that the second was the more expensive one.

For a suffering tooth, a filling is prescribed consisting of a mixture of frankincense, yellow ochre, and malachite.

It has often been emphasized that the recipes of the Ebers Papyrus are teeming with medicaments of the most repulsive kind. In fact, we find the dung of various animals—such as ass, sheep, pig, cat, gazelle, lizard, crocodile, pelican, and also fly's dirt (once with the cunning addition "that is on the wall") or dirt of the nail of a man and even urine and excrement of men sometimes among the ingredients of a salve, but here two facts must not be overlooked. First, that out of 877 recipes, only thirty-three (that is, not quite 4%) contain such repulsive ingredients. Secondly, that they usually appear at the end of a group of recipes prescribed for the same ailment, the majority of which are free from such ingredients. This seems to show that they belonged to a kind of folk medicine which was not in great esteem with the physicians themselves, but of which it was good to have some examples at hand in case a patient insisted upon their special value.

I do not know whether ingredients consisting of animal blood: blood of a calf, or an ox, of asses, pigs, hounds, and goats, and also of a bat or of various birds; or of the gall of animals (as ox, goat, tortoise) or birds or fish belong only in the sphere of folk medicine.

As to the gall of a special kind of fish, we are reminded of the pathetic story of Tobias, who cured his blind father with the gall of a fish.

As to the blood, we have a very clear indication of folk medicine, if we read that the blood of a black calf or the blood of the horn of a black ox is recommended for a salve against grizzling hair (other recipes prefer the backbone of a raven, a raven's egg, or fat of a black snake).

With these recipes against the grizzling of hair, and their sympathetic magic, we have come to a group of recipes in the Ebers Papyrus which, according to our conceptions, are not of a medical kind. Let me mention a few others: There are recipes to preserve the hair or to make it grow or to expel dandruff—but also to cause hair to fall out! The last one gives us an interesting insight into the life of an Egyptian harem. It was to be applied to the hair of a "hated woman"—this being the technical term for a fellow wife, who shared the master's favor with another! There are recipes to remove a thorn from the body, to improve the skin or to expel wrinkles of the face, and even recipes to prevent flies or gnats from biting, to expel fleas from the house or to prevent mice from approaching things. In the last case, the prescription is to cover these things with the fat of a cat.

The recipes of the Ebers Papyrus also have been discredited as being drenched with superstition and presupposing a situation in which, as Dr. Breasted expressed it, "magicians were contending with a demon-infested world." Dr. Ebbell has rightly emphasized in a brief preface to his translation that this conception is not correct.

We do not know how the ancient Egyptian physicians explained the origin of the numerous diseases which they treated. If anything is certain, however, it is the fact that they did not assume that all, or even a considerable number of them, were due to witchcraft of an enemy or to the influence of extra-human powers. In the whole Ebers Papyrus, the words "magic" and "bewitchment" occur just three times. "Magic in the belly" is mentioned twice, and four different potions are recommended against it. And a "remedy to expel bewitchment," which prescribes the swallowing of a big burned beetle, soaked in oil, is entirely unique and looks quite foreign within its surroundings. Magic and medicine always have existed—and do exist today—beside one another, and here evidently a text belonging to a book of magic has gone astray and got into a medical context.

Besides this, we find not more than eleven passages in which an ailment is said to have been caused by a deity or by a deceased person. It is interesting to see what kind of cases they are. Seven of them are cases of purulency and haematuria—one of them contained in one of the rare incantations of the Ebers, which I shall mention presently. One more occurs in a recipe against "afflictions" of a general nature which allegedly had been made by a goddess for the sungod Re. The three remaining ones—one against white spots in the eyes, the other against an ailment of the mamma, the third against cataracts in the eye—all occur within incantations!

These magical spells, however, are much rarer than Breasted's words would cause us to surmise. Besides the three that I have just mentioned, the whole Ebers Papyrus contains only ten more. They are prescribed against different ailments and diseases: one each against purulency and against blindness, one each in recipes to stop diarrhea, to kill the roundworm, against a fetid nose and against spotted baldness, two against a burn that putrifies, and one against a "swelling of vessels" in an evidently hopeless case, of which the physician is told, "Thou shalt not put thy hand to such a thing."

Taking them all into consideration, we receive the following impression: In some cases in which human help seemed to be impossible, a last attempt was made to get help from a supernatural source. In others, the remedy with incantations is added to a number of recipes without incantation—again for the physician to choose in case his patient absolutely wanted an incantation to go with the remedy!

If, finally, the text of the Ebers Papyrus is preceded by three incantations, which were supposed to be recited when applying a medicament, when drinking a potion, or when taking off a medical bandage, this can hardly be taken as anything but an expression in the belief of the assistance of the gods. We must not forget that even in the enlightened times of the Greeks, in the writings of Hippocrates, we find in a prominent place the invocation of Apollo and the oath to this patron of the physician.

The Edwin Smith Papyrus was bought in 1862 by an American of this name and has been published in 1930 in a masterly publication, including translation and extensive commentary, by the American Egyptologist James Henry Breasted, formerly of the University of Chicago.

In contrast to the Ebers Papyrus with its collection of the recipes for the *physician*, this Edwin Smith Papyrus represents —and is almost the only representative of—the second group of Egyptian medical texts, which were intended for the use of the *surgeon*.

It consists of a collection of forty-eight surgical cases, all built up according to the same realistically matter-of-fact scheme.

First comes a superscription, which briefly gives the name of the illness. This is followed by a careful description of the case in hand, which always begins with the words, "If you examine a man who" has this or that illness. Then comes a diagnosis which always begins with the words, "You should say" he suffers from this or that ailment. Here the same words reoccur in the main which were used in the first description of the case, though now, to be sure, as the responsible utterance of the attending surgeon. This diagnosis always ends with the words: "An ailment which" and then one of three possibilities follows. The surgeon may say: "An ailment which I am going to treat," or "An ailment which I shall combat," or "An ailment which cannot be treated." Thus, at the end of his detailed diagnosis, the surgeon must always give a kind of prognostic explanation. This prognosis is either favorable or doubtful or unfavorable. It is finally—except in entirely hopeless cases—followed by a method of treatment, which begins with the words, "You must do" this or that. Then the healing substances are given, of which I shall have more to say presently.

The cases of the Edwin Smith Papyrus are not only systematically constructed, each within itself, but their arrangement is throughout a systematic one.

The text begins with a group of injuries of the head or skull. These are followed by an injury of the forehead and the eyebrow, by nose injuries, injuries in the region of the cheek, of the temple or temple-bone, an ear injury, a fracture and a dislocation of the lower jaw, injuries to the upper lip, the chin and the neck, several collarbone injuries, some injuries to the upper arm, two wounds and a tumor on the chest, several injuries to chest and ribs, two cases of tumors on the chest, a shoulder injury, and finally, an injury to the lower spinal column.

The fact that here the text comes to an end would in itself make it probable that only a fragment of the original text has come down to us. The scribe who copied the Edwin Smith Papyrus from an older text actually has stopped in the middle of a sentence. We have every reason to assume that the complete text had once included injuries of the stomach, the pelvis, the leg and foot—perhaps also of the lower arm.

Even within these groups of cases, a systematic arrangement is noticeable, so that the author proceeds from simple head-wounds to more and more complicated fractures. Similar is the arrangement of the various injuries to the forehead and to the cervical vertebrae.

On one occasion, the cause of these injuries is indicated. It is said of the cervical vertebrae of a man, "His falling down headlong has caused the compression of one into the other."

We must think, here as well as in several other cases, of falls from scaffolding, which may have been frequent among the masons working on the Egyptian monuments with their often very considerable height. On the other hand, some of the injuries may have been inflicted by weapons, mainly by spears, clubs, and daggers.

The directions which are given for examining a patient refer, for example, to putting the hand on the wound and palpating it. Or the surgeon has to make certain observations, such as whether his patient can move his head sideways or down, whether he can open his mouth, move his neck, lift his arm, whether he is shivering, whether he bleeds (from nose and ears for example), whether the wound is deep, whether he has fever, whether the ends of a broken bone "crepitate," whether the patient can hear, etc. On one occasion, it is to be determined whether the pulse-beat is weak (the Egyptian says, whether "his heart is too tired to speak"). Several times it is prescribed that the physician demand certain motions from the patient, by saying, for instance, "Look at your shoulders," "Stretch out your legs, and put them together again," or that the surgeon makes the patient speak by asking him a question.

In therapeutics, the bandage is most commonly used. Of interest is the direction found in twenty-two, that is almost 50%, of the cases, that the patient be bandaged with fresh flesh. The idea seems to have been that the fresh flesh would stop the bleeding—just as today we treat haemophils with blood serum or even with fresh flesh. This flesh bandage is to be taken off after the first day, whereupon follows, in less serious cases, a treatment with fat and honey and often with a third undetermined ingredient of a vegetable character. This obviously was a salve applied with or without bandage, the constituents of which answer each its particular purpose. The fat remains moist and affords a protection against the outside. The honey —which generally takes the place of sugar in ancient Egypt— draws off the water and acts as an antiseptic. Whether the plant ingredient served the function of contracting the wound must as yet remain a conjecture.

Besides these more general prescriptions, there are a number of others for special cases. Several times it is prescribed that the patient be treated in a sitting posture: so in the case of skull fractures, fractures of the upper jaw, a temple fracture, a fracture of a cervical vertebra, and also in the case of an injured and swollen cheek. In particularly difficult cases in which the healing process would cover a long time, the exertion of sitting is to be lessened for the patient. "Make him two supports of bricks"—evidently for putting his arms on—is the direction we find in two cases of severe skull fractures.

In two cases of nose fractures, we find that tampons were employed, only they were soaked in fat instead of antiseptic, as with us. Three times it is demanded that the patient be treated in a reclining posture. First, in treating a collarbone injury; second, in the case of a fractured upper arm; and third, in a case dealing with a fracture of a thoracic vertebra without dislocation. The directions for treatment in this last case begin with the words: "You are to lay him on his back, and you are to prepare for him . . ." but what was to be prepared, we are not told. In the middle of the sentence, the scribe to whom we owe our text has interrupted his copying. It seems that he had intended to continue the text, for he left an empty space of sufficient length for four more columns, and after this gap, he began to copy a series of quite different texts. But he never fulfilled his intention.

One remarkable thing you will have noticed: that no surgical operation as yet has been mentioned. And, in fact, these ancient Egyptian surgeons do not seem to have employed the knife even once. But we do have one case in the Edwin Smith Papyrus where the surgeon operates, though not with a knife. It is the first of three cases which describe not injuries and fractures proper, but infected wounds and tumors. Concerning a tumor-like abscess, filled with pus, on the breast of the patient, the surgeon says toward the end of his diagnosis: "An ailment which I shall treat with the fire drill." And the author of the text goes on: "You should burn for him over his breast, i.e., over those tumors which are on his breast." The prescription here is, as we see, the burning out of a tumor—as is still done today under certain conditions with carbuncles, although with an instrument rather different from the primitive Egyptian fire drill!

In general, the atmosphere of sober observation and matter-of-fact practical prescriptions and treatments found in both of these outstanding medical texts of the Egyptians may be called scientific. This is especially true with the elaborate descriptions found in the surgical papyrus which refer to absolutely hopeless cases and could have none but a theoretical interest.

The Ebers Papyrus even contains two treatises, loosely attached at the end to the collection of recipes, which definitely show a systematic, scientific interest of these ancient physicians. One of them bears the title, "The physician's secret" or "The knowledge of the heart's movement and knowledge of the heart." According to this treatise, the human body contains forty-six so-called "vessels," which carry mucus and blood, sperm and urine, the breath of life and the breath of death. According to the other treatise, there are only twenty-two vessels in the human body, which lead to the heart and have connection with the limbs. They are arranged in 11 pairs, and they are of special interest for the physician because certain illnesses are supposed to originate within them. The importance of the heart for the entire body and its relation to the pulse-beat—the Egyptian says, "the heart speaks in all limbs" —is here correctly recognized.

While these first witnesses of a scientific spirit are common to both papyri, the differences between them are obvious. But these differences are caused by the difference of the subject matter treated. A compilation of recipes, prescribed against diseases, must be different from an enumeration of surgical cases.

On either side, the foundation is a sober and matter-of-fact attitude toward the various diseases as well as toward the various kinds of injuries, an attitude based on a long-standing practice and experience. To this attitude, we find, in the case of the recipes, a slight admixture of notions of folklore which go back to an earlier phase of civilization, and also of magical practices which may also be of an early origin, but which come to the fore again after the Middle Kingdom. Of these influences we have a very strange and isolated example even in the surgical papyrus. Here one of the cases—it deals with a complicated skull fracture—has neither a diagnosis nor a prognostic indication, and instead of the sober prescriptions of all the other cases, it gives a rather fantastic one and adds a magic spell exactly of the same kind as those found in the Ebers Papyrus. There can be no doubt about the later intrusion of the text of this case into its surroundings. The language proves that it belongs to a later time than the remaining text of the surgical papyrus. That the bulk of the main text goes back to the time of the Old Kingdom is shown by a great number of glosses added to the text of some of the cases, which explain words that in the course of time had become obsolete.

Thus we see that in Egyptian surgery exactly the same thing happened as in Egyptian medicine. At a certain time, the development of surgical knowledge and practice that was based on sober and matter-of-fact observation and experience, and had reached a remarkably high level, seems to have been stopped, and magical notions crept in, that were apt eventually to destroy what had been built up during the course of two millenniums. This is corroborated by the fact that the Egyptian surgeon, who once owned the roll which now we call the Edwin Smith Papyrus, did not insist, much to our regret, that the whole ancient surgical text be copied.

Instead, he had his scribe copy a number of quite different texts that were more to his liking and better fitting the general taste of his time. They are magic spells to be recited in order to "drive away the wind of a pestilential year" or "to drive away demons of sickness" or against the ill which a swallowed fly may cause in a man's body, and accurate directions for chanting them. They also contain a detailed recipe, which was supposed to have the power of changing an old man into a youth by means of a salve, the mixture of which is carefully described, and of which it is claimed that it "had been found effective many times."

This inclination toward magic is found much more developed in the three other medical papyri, which chronologically followed the Papyrus Ebers, one in Berlin, one at the University of California, and one other (the latest of all of them), in London. They all contain collections of recipes and magic spells.

It is possible that it was just the so-called Hyksos period, the time after the Middle Kingdom, from about 1800 to 1600 B.C., during which the change of views and methods in Egyptian medicine and surgery occurred. The proud structure of the Egyptian state had fallen down for the second time. The whole country was filled with unrest and uncertainty. It was the time when the Egyptians had lost confidence in their own strength and in their own achievements, and looked for help from the gods, who had forsaken them, because in the preceding times they had not served them as conscientiously as they should have done.

Illnesses and diseases were now supposed to be sent by the gods and demons, and through the help of gods and demons they had to be cured.

Thus we notice during the longer first part of Egyptian history, in medicine and surgery just as in other phases of civilization, a gradual development that reached a remarkable climax, probably about the height of the Middle Kingdom, ca. 2000 B.C.—while in the second, shorter part of Egyptian history, beginning about the time of the Hyksos invasion, a decline is noticeable.

UNIVERSITY OF PENNSYLVANIA
BICENTENNIAL CONFERENCE

Medieval Medicine

By

HENRY E. SIGERIST, M.D., D.Litt.*

We can approach the medical history of a period from different points of view, from that of practical achievements or from that of ideas. Medicine is a craft and a science. As a craft it is frequently transmitted by word of mouth and practical instruction, from father to son and from master to pupil. As a science medicine is one aspect of the general culture of a period. It reflects man's attitude toward nature, toward the phenomena of life and death. It is expressed in literary form, and the medical books represent one aspect of the literature of a period sharing its general style. We may be more interested in the health conditions and health hazards of a period and in the treatments and diets applied to cure disease or to prevent it. Or, we may be more attracted by the ideas that guided the physicians' actions. In the following brief sketch of medieval medicine I shall not be able to discuss its practical attainments in a more than cursory way. Rather, I will try to determine the place of medieval medicine in the history of civilization.

Our symposium has a serious gap in that it jumps from ancient Egyptian to medieval medicine and has omitted a discussion of Greek medicine. And yet it was Greek experience and Greek thought that constituted the basic content of medieval medicine. Greek medicine was transmitted to the medieval world and was gradually assimilated by it. A synthesis of rare harmony was achieved between Greek and medieval views until, in the Renaissance, the Western world revolted against traditions. Let us examine this process.

*William H. Welch Professor of the History of Medicine and Director of the Institute of the History of Medicine, The Johns Hopkins University.

TRANSMISSION

We must remember that medieval medicine had two centers of development, the Muslim empire in the East and the Christian world in the West.[1]

In the seventh century A.D., Arab tribes, driven from their homeland by the aridity of the soil, united and disciplined by a new creed, moved north seeking more fertile lands. They conquered Syria, turned west, conquered Egypt, the whole coast of North Africa, went over to Spain, crossed the Pyrenees, until they were stopped in France. In less than a century they had founded an empire that reached from the Pyrenees to the Indus River. They were tolerant. No one was forced to embrace the new religion, but the infidel was a subject, heavily taxed. It was highly profitable to become a Mohammedan. The convert acquired Arabic citizenship and became a member of the ruling class. Millions adopted a religion which, after all, was not so different from Christianity. The new empire was united by a common faith, disciplined by religious rites and, since the Koran was not to be translated, it had a common language.

The Arabic conquerors were a rough crowd, horsemen, warriors, poets at times, but little experienced in the arts and crafts. They soon found that the people they had subjugated had better architects, painters, engineers, and physicians. They hired them and soon began learning from them. Alexandria, although the famous library had been destroyed before the conquest, was still a center of learning. It was obscured by mystic currents but was backed by a great tradition; Paulus of Aegina, the last great Greek medical compilator, lived there in the first half of the seventh century.

More important because it was infinitely more dynamic was another intellectual center, Gondeshapur, in Persia. A foundation of Sassanian kings, it had become an asylum for refugee scholars. Greek philosophers driven from Athens by Justinian, Christian heretics, Nestorians, driven from Nisibis, convened in Gondeshapur where they came in touch with Persian and Indian thought. Of all the sciences medicine was probably the

[1] If I speak of East and West, I do it for the sake of brevity. I am well aware that throughout the Middle Ages there were Mohammedans in Spain and Christians in Syria.

most flourishing, centered around a hospital and an academy. Syriac had become the language of learning in Western Asia, and many classics of Greek philosophy, science, and medicine were translated into Syriac.

And so the Arabs found in the territory of their conquest not only intellectual centers, but they found in addition the chief Greek medical literature already translated into a Semitic language. Books were the source of knowledge, and the delivery of books, particularly of alchemical and medical books, was more than once made a condition of peace treaties with the Byzantine empire. Once a book was available in Syriac version it was an easy matter to have it translated into Arabic. After this was done, the book could be read and used by all who needed it from the Pyrenees to India.

In the second half of the eighth and throughout the ninth century an endless number of Greek books were translated in this way: the works of Galen and his successors, but also Hippocratic writings and the Materia Medica of Dioscorides, the latter gorgeously illustrated and a book that is still consulted in the Orient today. The chief "transmitter" was Hunayn ibn Ishaq, who was head of a regular school of translators in Baghdad, at the court of the Abbasid caliphs. He was assisted by his son Ishaq and his nephew Hubaysh, and tradition attributes to him over ninety pupils. Most of these translators were Christian scholars. They were the linguists of the day, mastering Greek, Syriac, Arabic, and often Persian. They usually first translated a book into Syriac for the use of their fellow Christians, then into Arabic for the use of Muslims. Just as the Ptolemies in the third century B.C. had sent out regular expeditions in search of Greek manuscripts for the Alexandrian library, so did the Abbasid caliphs for the library in Baghdad.

At the end of the ninth century the Arabic-speaking world was in full possession of the Greek medical tradition. It was a medical science that had lost its momentum and had completed its course. It was, moreover, the science of another people, different in race and outlook. Nevertheless, it was the accumulated experience of centuries of observation and reasoning, innumerable facts about diseases and their treatment, that became available to the Islamic world in this way.

Developments were similar and yet different in the West. There too it was hunger that drove barbaric tribes into the fertile fields of the Roman empire and started a migration of nations. The Germanic people that settled in the Western part of the Roman empire had just as primitive medical knowledge as the Bedouins of Arabia. They too all of a sudden came in touch with a much higher civilization. In the East the Arabs went on their conquest with a new religion that was gradually adopted by the subjugated nations. With their religion they took over the conqueror's language, at least as a literary language. In the West the process was different. The Goths were converted to Christianity, which was the official religion of the Roman empire. Its language was Latin, and Latin became for centuries the literary language of all nations that recognized the authority of the Church of Rome.

For over one thousand years medical books had been written in Greek throughout the Graeco-Roman world, and very few Latin medical books were available. But Latin was becoming an increasingly important language. It was the language of the court, the administration, and the church. It was the vernacular language of Italy, Gaul, and Spain, and was the literary language not only in Germany but in North Africa and Britain. There was a strong demand for medical books written in Latin. Translations were made in the West as in the East, and from the fourth century on a new medical literature developed which was written directly in Latin. It was not original in character, but consisted of compilations, its value depending on the sources used. The West had intellectual centers also. One such center was North Africa, in the fourth and fifth centuries, where Saint Augustine lived and where one of his friends, the physician Vindician, and his pupil Theodorus Priscianus compiled some important books. An African, Caelius Aurelianus, translated Soranus and thus preserved the experience of the Methodist school, the doctrine of which was very influential in the early Middle Ages.

Another such center was Bordeaux, famous for its school of rhetoric. But there were physicians there too, such as Marcellus Burdigalensis, who compiled a very popular collection of prescriptions. Ravenna, the residence of Theodoric, had in the sixth century a medical school, the outlines of which we just begin to perceive. It had Iatrosophistae, professors of medicine, who interpreted the Galenic canon in Latin in the same way as was done in Alexandria in Greek. Oribasius was translated twice in Ravenna, and there can be no doubt as to the importance of this school. Roman institutions did not perish in the early Middle Ages. Many schools flourished in Italy in the Lombard period and became starting points of universities, as was the case in Bologna.

At the time when Harun al-Rashid attracted scholars and artists to his court, Charlemagne did the same in the West and laid the foundation for schools that were to become famous, in Tours, Chartres, Rheims. At the same time Benedictine abbeys, like Monte Cassino, Bobbio, St. Gall, Fulda, were centers of learning where ancient literature was studied, copied, and passed on from generation to generation.

There can be no doubt, however, that the Muslim world was far ahead of the Western world in the early Middle Ages. Around 900 the Arabs were in full possession of the Greek medical tradition. In the West some works of Hippocrates and Galen, the Materia Medica of Dioscorides, Soranus and some other great writers were translated as early as the sixth century, but around 900 they were almost forgotten. The popular literature consisted of short treatises compiled for practical purposes in Greek in the fourth century mostly, translated into Latin in the sixth century. They were translated into Syriac, Arabic, and Hebrew also, but were superseded by better literature. In the West, however, these short treatises dealing with urine, pulse, fever, diets, prognostic, bloodletting, and pharmacology constituted the bulk of ancient literature that was still alive in Carolingian days.

Dioscorides, the chief source of ancient materia medica, was translated three times in the early Middle Ages, but of one version we have only an indirect testimony. A second version is preserved only in short fragments, and of the third only two manuscripts have survived, while we still have over fifty manuscripts of the herbal of Pseudo-Apuleius. This shows how infinitely more popular this very inferior treatise was. The prognostic of Hippocrates was translated twice, but both versions are known only in short fragments while there are many manuscripts of the so-called *Prognostica Democriti*.

In 732 the Arabs were repulsed from France, but they remained in Spain until 1492. They conquered Sicily in 827 and ruled the island until the end of the eleventh century. From the eleventh to the thirteenth century East and West clashed in the crusades. Intercourse between the two civilizations became very close. Different as they were, they had a great deal in common due to their common heritage. Not only commercial but also intellectual relations increased and, since the Arabs were more advanced in science and medicine, Europe began to learn from them.

In the eleventh century, Constantine, an African by birth and therefore a master in Eastern languages, traveled all over the Orient and came to Monte Cassino where he became a monk, bringing with him Arabic medical books that he translated into Latin. He thus greatly enriched Western literature and made Greek and Arabic writers available that had not been known before.

Toledo, one of the chief centers and the Western outpost of Arabic learning, was conquered by Alfonso VI of Castile in 1085, but it maintained its position under Christian rule. It thus became the center from which Eastern knowledge was transmitted to the West. In Toledo in the twelfth century Gerard of Cremona and his students translated a large number of Greek and Arabic medical, scientific, and philosophic writers from Arabic into Latin. In the early thirteenth century the Western world possessed the Greek medical tradition as the Arabs had done three hundred years before, and possessed in addition the experience of many Arabic scholars.

ASSIMILATION

The Greek medicine that was transmitted to the Middle Ages, in the East and in the West, was the result of a development of over one thousand years. All schools of thought from the early pre-Socratic philosophers to Plato, Aristotle, the Stoics, Epicureans, and Skeptics, to the vagaries of Neo-Platonists and Neo-Pythagoreans, were reflected in some way or other in the physicians' theories. This enormous mass of literature was transmitted in a relatively short period of time, in a haphazard way without any order. A book was translated when good manuscripts were available. This determined the choice first of all. Hunayn ibn Ishaq translated one day the Hippocratic Aphorisms, a book written around 400 B.C., and some other day he made a version of a Galenic treatise written almost six hundred years later. Gerard of Cremona within a few years rendered such disparate books in Latin as the *Techne iatrike* of Galen and the *Liber Almansorius of Rhazes.* This new literature was taken over, not as a collection of historical documents, but as a living whole. It was studied not by medical historians, but by physicians desirous of learning from it how to treat their patients and how to comprehend the phenomena of health and disease. Viewed from such an angle, the Greek medical tradition was extremely bewildering. It was full of contradictions. Many descriptions of diseases and many prescriptions were unintelligible. A theory that a Greek physician familiar with the philosophy of Pythagoras found easy to understand seemed strange and foreign to an Arab or Christian cleric of the early days. A great deal of interpretation was required before this new learning could be assimilated.

Dictionaries were written for the elucidation of difficult terms or concepts, commentaries to explain authoritative texts, concordances in which similar opinions were brought together, conciliatores to reconcile divergent views. Such books were written in the East and in the West.

The Greek tradition, however, carried to the Middle Ages not only doctrines but basic observations and methods. It taught that disease is a natural process not essentially different from physiological processes. It taught further that the human body has a natural healing power which tends to overcome lesions and to restore the lost balance of health, that all actions of the physician must therefore be directed toward aiding this *vis medicatrix naturae.* The Greek tradition taken as a whole regardless of doctrines taught how to approach a sick man, what questions he should be asked, how to examine him, and how his symptoms must be evaluated so as to know what fate has in store for him. Greek medical literature of all periods was full of unsurpassed descriptions of disease symptoms and disease pictures. And it contained a wealth of information concerning the treatment of diseases—dietetic, pharmacological, physical, and surgical—the result of centuries of experience. Once this knowledge was assimilated medicine could advance. And it did, in the East and in the West.

The tenth and eleventh centuries were the Golden Age of Arabic medicine. The leading physicians were no longer Christians but Muslims. They came from all parts of the empire, many of them from Persia. Hospitals were built in increasing numbers from the ninth century on. They were not poorhouses or almshouses like the Western hospitals of that period. They were places where sick people were treated, where physicians gathered experience and instructed students.

The number of Arabic-writing physicians who enriched medical knowledge is large. Many of their writings are lost or still buried in manuscripts. Let me mention only a few names and a few contributions. Al-Razi (Rhazes), probably the greatest Muslim clinician, was an extremely versatile scholar, physician, scientist, philosopher, and theologian. We admire him not so much for his *Continens*, an encyclopedic textbook of medicine, as for his case histories, monographs, and short treatises in which he established new disease entities. Most famous is his book *On Smallpox and Measles*, remarkable also his treatise *On Stone in Bladder and Kidneys*. Many more are still unpublished. Rhazes' medical doctrine was Greek, to be sure, but by applying Greek methods of clinical observation and research he enriched medicine considerably.

Another distinguished clinician of the period was Ali ibn el-Abbas (Haly Abbas), like Rhazes a Persian. He too wrote a comprehensive textbook of medicine which is full of valuable observations and reflections. He took a critical attitude toward his predecessors, Greek and Arabic, and accepted from them what he considered true.

All sections of the empire contributed to the Golden Age of Arabic medicine. An Egyptian Jew, Isaac, wrote important monographs on fever, urine, diets, and drugs. One of his students, Ibn al-Jazzar, became well known for a little book in which he gave dietetic advice to travelers. It was translated not only into Latin but also into Greek and Hebrew. The greatest surgeon of the period, Abu'l Kasim, was born in Spain, in El-Zahra near Cordova. He was influenced by Greek writers, notably Paulus of Aegina, but was an experienced surgeon himself. Materia medica was greatly enriched by Arabic writers. An empire that covered such a vast territory yielded drugs from all climates. A mere list of Muslim physicians who contributed to the subject would fill many pages and include names from all provinces.

The Greek tradition was assimilated in the West also, but later than in the East and in a somewhat different way. As we mentioned before, it was transmitted, not in its pure form, but after having gone through the Arabic channel. The experience of the Greeks was made available to the Western Middle Ages together with that of the Arabs. Constantinus Africanus in the eleventh century translated not only works of Hippocrates and Galen but also those of Rhazes, Isaac Judaeus, Ibn al-Jazzar, and other Arabic writers.

Constantinus' work marks a turning point in Western medieval medicine. It became known in Southern Italy just at the time when the School of Salerno was developing vigorously. Salerno was a trading town where Greek was heard in the streets and where Western and Eastern influences converged. A group of physicians, laymen and clerics, were practising in the town, sought by patients and by students from all over Europe. In response to a strong demand for a richer medical literature they compiled books such as the *Passionarius Galeni* which, however, still had all the characteristics of the early medieval literature. The translations of Constantine acted as a strong stimulus. They found in Salerno a group of physicians that was ready to absorb and assimilate them. The literature that Salerno produced in the twelfth century started a new movement in Western medicine. The many books they wrote on all subjects of practical and theoretical medicine reveal that the Salernitan masters had not only assimilated the Graeco-Arabic tradition but had already been able to add observations of their own. It is highly significant that they were fully aware of the importance of anatomical studies. Human bodies were not yet dissected, but those of animals were.

Another important contribution to medicine, though of a different order, came from Southern Italy. Frederick II in his *Constitutiones* of 1240 set definite standards for the practice of medicine by requiring a prescribed curriculum of nine years, examination by the Salernitan masters in the presence of a representative of the state, and by licensing the medical profession. This gave it a status it had not had before.

When Gerard of Cremona and his group were at work in Toledo, another medical school had come into existence not far from Spain, in Montpellier. Just as Salerno had profited by the first wave of translations, Montpellier did by the second. The interpretation and assimilation of this new literature became one of the chief tasks of the young Western universities.

If we wish to watch the medieval physician at work, we must not only consult the textbooks. Textbooks, even in our days, always have to a certain extent the character of compilations, since no man's original researches can cover an entire field. We must read the *Consilia*, missives in which a doctor discussed a definite case. Or we must watch him fighting epidemics. When the Black Death ravaged Europe in 1348, the physicians had to face a problem for which ancient medicine did not give any solution. Or we must look at the surgeon operating on a soldier after a battle.

When we do this, we soon find that Western medicine too had absorbed the Greek tradition and was enriching it by many important observations.

SYNTHESIS

So far we have spoken of the transmission of Greek medicine and its assimilation and enrichment by medieval physicians. Was this all? Was medieval medicine nothing else but a reminiscence of ancient Greece, a belated outgrowth of Hellenistic medicine? Is it possible for a civilization that is alive to take over ideas and systems which are deeply rooted in another civilization without modifying them? The Middle Ages, in the East and West, produced new forms of expression in the social and economic life, in government, law, theology, art, and literature. Is it conceivable that they could have left medicine without their imprint? In other words, is there such a thing as an essentially medieval medicine?

Of course there is. A synthesis was accomplished in this field also. So far, little research has been done on the subject and all I can do is to show where this synthesis is to be found.

A work like the *Canon* of Avicenna could not have been written in antiquity. Avicenna, one of the greatest physicians and philosophers of Islam, attempted to build a complete, logical, and well-rounded system of medicine. Its elements are to a large extent Greek—Greek medical experience and thought, Aristotelian philosophy, with a tinge of Neo-Platonism. To this was added the experience of several centuries of Arabic medicine and a great deal of personal experience. With these elements in hand, Avicenna created a system that was no longer Greek but was an expression of Muslim philosophy. It was so forceful and persuasive that it dominated medicine in the East and the West for six hundred years.

In another sphere of medieval culture we find a physician-philosopher, Maimonides, who wrote *Aphorisms according to Galen*. The book is by no means a mere repetition of Galenic doctrines. Maimonides selected passages from Galen. He selected what appealed to him particularly, and the choice he made already reflected his personality. He took a statement from Galen as motif and developed it in his own way, thus creating a synthesis of Greek, Arabic, and Jewish thought.

The same synthetic process can be traced in the works of the Western scholastic physicians of the thirteenth and fourteenth centuries, Albertus Magnus, Roger Bacon, Arnald of Villanova, Pietro d'Abano, to mention only a few. Aristotle, Galen, and Avicenna were their masters. They quote them constantly and follow their methods. But they did more. They were Christian scholars. Theology was the mother of science and learning, and they succeeded in creating systems in which the experience of medicine became part of the Catholic concept of the world. Their works are essentially medieval.

REVOLT

The parallelism in the development of medicine in the East and in the West is striking but is easily explained by the common heritage and by the whole situation in which both groups of people found themselves in the early Middle Ages. It is much more difficult to explain why this parallelism came to an end.

The Golden Age af Arabic medicine was short. After 1100 there was a steady decline. Factual contributions to medicine were still made and many books were written, but there was hardly any development. People looked backward and not into the future, commented upon their classics and followed traditional patterns of thought. The Islamic world remained medieval to our days, except in the few sections that have recently adopted features of Western civilization.

Matters were different in the West, and the Renaissance marked the turning point. It is a matter of speculation to determine what forces created that great and deep movement. I shall not attempt to discuss the problem in this brief paper. There was a primitive accumulation of capital in the East just as much as in the West, perhaps even more, but it was Europe that developed a capitalist economy. Great voyages of discovery were undertaken by the Arabs long before Europe was thinking of a sea route to India, but the European voyages had a much more profound influence. They affected Western economy deeply and became a stirring experience.

One of the essential traits of the Renaissance was its attitude of revolt against the traditional authorities. The most powerful medieval authority, the church, was attacked and "reformed." The power of the craftsmen's guilds was broken by the developing industry. The authority of the medical faculties was opposed, and their power to regulate the practice of medicine was gradually taken over by other agencies.

Throughout the Middle Ages the Greek medical tradition was accepted as authoritative. It was open to interpretation, to be sure, but its authority was hardly ever questioned. Now physicians wrote books *De Plinii et aliorum medicorum erroribus*. This revolt against tradition was sometimes dramatic, as in the case of Paracelsus. It was usually less spectacular but was a revolt nevertheless, and it paved the way to a new medical science.

UNIVERSITY OF PENNSYLVANIA
BICENTENNIAL CONFERENCE

The Rise of Modern Scientific Medicine

By

RICHARD H. SHRYOCK, PH.D.*

THERE is a sense in which the scientific status of medicine has been less secure during the past two hundred years than it was during earlier centuries. Galilei turned from physical to medical research without any apparent feeling that the latter was less "scientific" than the former. Indeed, medicine had long been viewed as the scientific field *par excellence*. In the eighteenth century, however, one encounters the view that this field was failing to keep pace with the physical sciences. By about 1840, such opinion had become so common that a speaker before the British Association for the Advancement of Science openly referred to medicine as "the withered arm of science." And even in our own day, we are occasionally reminded by thoughtful observers that this or that branch of medical investigation cannot yet be viewed as a "true science." That such skepticism concerning medicine should have developed in the very period of its greatest progress obviously measures not a decline in medicine, but rather the development of more exacting standards in defining the nature of science.

There has been a tendency, with which we are all familiar, to deny the scientific character of any research which does not depend primarily upon quantitative procedures, and even to limit the "true sciences" to those disciplines capable of prediction within the limits of a not-too-disturbing margin of error. If we accept such definitions and then assume that "modern medicine" is synonymous with "scientific medicine," one might hold that the modern era in this general field is only dawning on us at the present time. Yet such a view would seem to obscure some of the most significant developments in the long history of medicine.

The idea that the essence of science is the ability to predict, whatever its philosophical presuppositions, is a pragmatic one

*Professor of American History, University of Pennsylvania.

emphasizing results. Paralleling this conception, there has been a similarly pragmatic popular view as to the criteria for modernity in medicine. This is the general assumption that "modern medicine" arrived when it became possible to prevent or cure disease on a considerable scale. Measured by this criterion, modern medicine began largely with the age of Pasteur and Koch. This view is widely held by practising physicians as well as by laymen, and its popularity is attested at this very time by no end of contemporary "movies" and best sellers.

Here again is a philosophy which obscures those fundamental changes that mark the advent of truly modern medicine. No one has made, nor is ever likely to make, a popular "movie" about a Morgagni or a Bichat. Such men, if any, deserve the title of the founders of modern medicine. But they could not predict and they did little in the way of cures. Much of their best work was done in the death house, and autopsies do not provide promising material for Hollywood. On what grounds, despite all this, may we identify the advent of modern medicine with their period—let us say 1750 to 1850—and especially with the last half-century of that epoch?

It will be assumed here that the essence of science, speaking in very general terms, lies in the attempt to observe and interpret phenomena rationally and in terms of all means to verification that seem available. Assuming the particular philosophical approach to science which was well established by the seventeenth century, the degree of success attained in any field at any given time was largely determined by the effectiveness of the methods of observation and verification which could be employed. Individual discoveries of practical import might be made by so simple a method as trial and error, or even by accident—as is seen in the occasionally valuable contributions of folk medicine—but no continuous or extensive development of either so-called pure science or its technological applications proved possible without more adequate procedures. One of the most significant themes in the history of any science, therefore, relates to the development of methods, in both their philosophical and technical aspects, since these determined so much of what followed in the way of discovery and its application.

It is a truism that the most effective methods employed in modern science have been those calculated to promote more exact and more searching observation; that is, the use of instruments to aid the senses, of quantitative procedures, and of experimentation. The slow development of these methods in physics and technology through the later Middle Ages, leading up to their more successful application in the sixteenth and seventeenth centuries, is a well-known story. Their systematic employment is associated first with the age of Galilei, and it is no accident that we commonly date the advent of modern physics at that time. The same methods were then employed in physiological and bacteriological studies, yet we do not therefore view this era as marking the beginning of modern medicine.

The truth is that if we think of medicine as representing a group of sciences focused upon the problems of sickness and health, physiology and bacteriology were not strictly medical sciences but rather biological disciplines of potential value only to some future medicine. Medicine proper, in terms of the science and practice of even the leading physicians, was largely unaffected by modern methodology throughout the greater part of the eighteenth century. Elaborate hypotheses were unchecked by experiment, microscopes were neglected, and even such simple quantitative procedures as measuring the pulse by a time-piece were commonly ignored. This failure to take over promptly the methods of the physical sciences was partly due to difficulties in developing instruments, partly to the complicated nature of the phenomena with which physicians had to deal.

Perhaps more serious than methodological difficulties, however, was the failure to envisage and concentrate on the essential problem whose solution was a *sine qua non* for further progress in the medical sciences. Physicians were misled in this matter by their humane interests and by the social pressure of popular demand—a type of pressure to which physical scientists were rarely subjected. It was naturally assumed that the main problem facing medical scientists was to cure illness, when the truth was that this desideratum was itself contingent upon the preliminary solution of a still more basic question: What was the nature of illness itself? It was as if one could not exploit a term without first defining it.

After all, medicine was indeed a group of sciences and related arts focused on the one major phenomenon of disease. There could be sciences of anatomy and physiology, as branches of biology, in a world without disease—but certainly no medicine. The primary problem in medicine, we can now see, was therefore abnormal rather than normal anatomy and physiology. The latter were essential only as a means of comparison—as the limiting aspect of the abnormal. Of course physicians had long thought about the abnormal both as a condition and as a process. But they had difficulty in distinguishing the importance of these two concepts, frequently viewing the relatively meaningless congenital abnormalities like monstrosities as of equal significance with disease processes. More unfortunate was the tendency to view even abnormal processes as simple, subjective conditions to be differentiated only in terms of the parts affected—as, for instance, illness of the head, of the chest, and so on; or, more usually, as illness either of the solid or the fluid parts of the body. Those who accepted the theory of a fluid or humoral pathology talked vaguely of vapors and impurities; while those who insisted just as dogmatically on solidistic theories lectured learnedly of tenseness and laxity. The temptation to maintain such "systems" of pathology lay not only in their logical simplicity, but also in the fact that they led obviously to a similar simplicity in therapeutics. One cause or condition of all illness implied one cure, and the doctor who followed this logic was likely to view himself and to be viewed as the benefactor of mankind. Thus John Brown of Edinburgh proclaimed, about 1790, that all illnesses were due either to tenseness or to laxity in the solid parts, and that all could therefore be cured or improved by soporifics and stimulants—by laudanum and Scotch whiskey. This system added to the appeal of logical simplicity a certain personal fascination which proved irresistible in some circles.

As long as all the complicated phenomena of illness were thought of in such simple, undifferentiated terms, even good methods of research were of small value. There was really no need for them—everything was settled. Note that it was the problem on which eighteenth-century physicians focused which thus nullified the newer methods with which some of them actually were familiar but rarely employed. The question of cures, we now know, was really too complicated to permit of quick results by the best of methods, and they therefore ignored the latter and chose rather to follow speculative short-cuts which brought the appearance of solutions.

That such speculative medicine was still dominant at the end of the eighteenth century was due not only to the natural demand for cures—for focusing on the problems of the moment—but also to the apparent failure of other approaches. During the seventeenth century a number of medical leaders, notably the Englishman Sydenham, had actually raised the problem of the nature of illness, which they felt should be differentiated before there could be any differential and therefore effective treatments for the same. They began to think in terms of diseases rather than of illness in general. The idea was not entirely new. Certain specific disorders with striking superficial symptoms like measles and smallpox had been identified during the Middle Ages. Nevertheless Sydenham emphasized the concept of distinct disease entities, and the idea took hold.

Unfortunately the only basis upon which diseases could then be distinguished was that of symptoms, and symptoms were so numerous as to be endlessly misleading. By the later eighteenth century, the published nosographies listed as many as eighteen or nineteen hundred supposedly separate diseases identified with as many symptoms or symptom-complexes. One suspects that the nosographers yielded to the lure of classification as such which was characteristic of the age. The process worked well enough in botany, where one dealt with tangible species, but was another matter when one played on paper with the names of human symptoms. It was in part a protest against the resulting confusion which led impetuous men like Brown of Edinburgh and Rush of Philadelphia to go to the other extreme with their doctrine of only one basic condition or disease.

The more level headed doctors who tried to follow Sydenham in distinguishing between one disease and another tended in practice to avoid the over-complicated and impractical nosographies. They did not go to the other extreme of a monistic pathology like that of Brown, but they did have to think in terms of a few over-simplified "clinical pictures" like "dropsy," "inflammation of the lungs," "fevers," and so on. These concepts were not clear-cut enough to encourage a search for specific causes or cures. It was indeed observed that "fevers" in general were increased in some mysterious way by insanitary living conditions, and much more was done to prevent fevers by sanitary reform during the first half of the nineteenth century than is now usually recalled. Such sanitarians as Southwood Smith of London, Lemuel Shattuck of Boston, and the almost unknown Wilson Jewell of Philadelphia deserve far more recognition in this connection than they have usually been accorded. But the fact remains that the sanitarian's approach was largely an empirical one as long as he did not know the exact diseases with which he was dealing, and this set serious limits to his achievements.

Even when a cure which we now know to be specific was discovered more or less accidentally in folk medicine—as in the case of cinchona—its effectiveness was limited by the confusion of "fevers" to which it was applied. We now know it helped with malaria but was useless with other infections not then identified. Conversely, it will be recalled that the one disease to which the preventive practices of inoculation and vaccination were successfully applied during the eighteenth century—smallpox—was one of the few conditions then identified as a distinct entity.

In retrospect, then, one may say that the grand medical problem of that age was to follow Sydenham's lead in identifying diseases through symptoms, but to refine it in such a way that the entities defined be neither too generalized, like those just noted, nor too detailed, like those of the nosographies. This problem was solved by a remarkable group of research leaders who found a clue in the correlation of two research trends which had hitherto followed more or less independent paths. This was the correlation of clinical descriptions with pathological anatomy—of the bedside findings with those of the autopsy. For centuries anatomists had pursued their own way, often out of pure biological curiosity, and had gradually acquired some knowledge of pathological processes. At the same time, they slowly attained the view that there might be some connection between the antemortem behavior of the body and the postmortem conditions which it exhibited.

This view was first effectively expressed in the classic work of Morgagni published about 1760, whose lead was later followed by the great Paris school of clinicians and pathologists. The French leaders from Bichet to Laennec and Louis—1800 to 1840—patiently followed hundreds of cases through hospital wards to the death house, and systematically checked symptoms against lesions, until a new picture of disease entities began to emerge. It should be noted, in passing, that such investigations would not have been possible except in large hospitals, and that the development of these institutions between 1750 and 1850—

itself a function of the growth of great cities—was therefore as essential a condition to the evolution of modern medicine as was any particular method or approach. This dependence of modern medicine upon the "culture of cities" well illustrates the intimate association of social with technical factors in bringing about the scientific developments of recent centuries. Astronomy could be, indeed has to be, pursued in rural retreats; medicine requires just the opposite setting.

As a result of such research as was done at Paris, the older humoral pathology was frequently found to be meaningless—diseases were often clearly associated with lesions in organs or within the tissues of the same. Moreover, the less extreme but still too generalized pictures like "fevers" or "inflammations" could be differentiated as hiding a number of more definite conditions corresponding to specific lesions. The "clinical picture" of "inflammation of the lungs" could be shown at times to correlate with the existence of distinct tubercles in the lungs, at other times with a congestion or consolidation in the lung tissue, again with a congestion in the bronchi, or in the pleura, and so on. In a word, there emerged the concept of such distinct diseases as phthisis, pneumonia, bronchitis, and pleurisy. All this took time, and meanwhile the pathological research had to be carried on by apparently cold-blooded men who resisted the siren call for immediate cures. For them, it seemed, death was but a scientific fact; and they even went so far as to develop a general skepticism about the possibility of cures. No wonder that the laity were discouraged. Imagine the feelings of patients and of conservative doctors as well, when Oliver Wendell Holmes, trained in the Paris school, declared that if most of the medicines in America were cast into the Atlantic it would be so much better for mankind and so much worse for the fishes! Yet these very cold-blooded pathologists, these "nihilists," were preparing the only basis upon which a succeeding generation could at last build a systematic structure of prevention and cure. For no one, save by blind trial and error, could seek the causes or the cures for a disease until that disease was itself identified. Hence we have the paradox that the very period of 1830-1870, in which the public became so discouraged about the services of physicians, was actually the most revolutionary and promising one in the whole history of the medical sciences.

Just because the public did become discouraged by medical

nihilism they tended to turn in these years to medical sects which promised all things. Actually, these sects represented survivals of the older speculative pathologies, with their extravagant theories that all diseases were really simply this and that, and that all could therefore be cured thus and so. Hahnemann, Thompson, Still, Mary Baker Eddy, and their type were all the logical descendants of John Brown and Benjamin Rush—each with his all-pervading cause and his all-sufficing cure. Some of these sects had real merits. But the rejection of their systems by regular medicine after 1840 in itself indicates the triumph of a critical and realistic attitude which may well be taken as the criterion for the advent of modern medicine.

The actual process by which the clinical-pathological research of the early nineteenth century led to the therapeutic achievements which followed after 1870 takes one back to the consideration of methods. It will be noted that modern methods played little or no part in beginning the reorientation of pathology early in the century. Bichat had small use for instruments of observation, for quantitative procedures, or for experimental verification, as he went back and forth from the wards to the autopsy table. But the beauty of this new approach, of this focusing on the really significant problem, was that it soon led to a need for improved methods—for techniques that would never have been called for if the older speculative pathology had continued to prevail. Then, by a familiar process, the need for instruments or techniques led sooner or later to their discovery and improvement. Or, if already available because of advances in other fields, older methods were now taken over and applied in medicine instead of being ignored as heretofore. Laennec, desirous of more exact observations in the wards, recalled and systematically applied the neglected stethoscope. Other "scopes" were added during the next generation. One of these, the older microscope, was vastly improved by the invention of achromatic equipment. Louis, anxious to check more exactly the results of treatment in the now identified pneumonia, bethought him of the use of a rapidly developing branch of mathematics—statistics—and others refined this by employing also the calculus of probabilities. The use of clinical statistics, as well as that of the various "scopes," soon became an established technique in medical research. Sooner or later, also, experimental verification of hypotheses in pathology was introduced, largely through animal experimentation. It need hardly be added that the whole development of pathological anatomy was paralleled, somewhat later, by similar progress in pathological physiology, and that all the types of method just noted were employed therein.

The relative place or rôle of those arts and sciences which the public especially identifies with modern medicine—notably of medical bacteriology and of surgery—now becomes apparent. Both the general procedures and even the essential ideas exploited by medical bacteriologists were provided by the anatomists and physiologists of the first half of the nineteenth century. It may be repeated, too, that medical bacteriology would have stood still—as it did during the eighteenth century—had not the pathologists identified diseases and thereby provided bacteriologists with something on which to work. One of the greatest services performed by medical bacteriology after 1870, incidentally, was to render still more exact the identification of distinct diseases. This it did by adding to the earlier criteria of symptoms and lesions the major criterion of cause. Just as a condition known only through symptoms could be more clearly conceived if it was also defined in pathological terms, so disease conditions based on these clinical and pathological bases could be still more sharply defined if a specific cause could be found. Thus the clinicians gave us "inflammation of the lungs," the pathologists narrowed this down to pleurisy or pneumonia, and the bacteriologists finally divided the latter concept into "types 1, 2, 3," etc., in terms of the causative organisms involved. The dramatic achievements in prevention and cure which followed have sometimes obscured the essential value of this preliminary phase of bacteriological achievement. The basic importance of the latter contribution may be recalled very easily, if one considers the great difficulty now experienced in dealing with illness that still defies exact identification. Thus "the common cold" is still pretty much a complex "clinical picture," as little is known definitely about either lesions or causes associated with it. It is thus still in the same category with the old "fevers," and consequently may conceal various diseases with which it is now very difficult to deal except in an empirical manner.

The development of modern surgery during the last half of the nineteenth century is ordinarily credited to the impact on that field of chemical and bacteriological discoveries; that is, to the introduction of anesthetics and of aseptic procedures. These developments, however, might well be interpreted as a consequence rather than as a cause of modern surgery. As a matter of fact, certain anesthetics were well known to chemists for a generation or two before modern surgery appeared, but no one bothered to use them. Was it not, as Sigerist has pointed out, the new pathology which really revolutionized surgery? This field was largely limited to emergencies like amputations and to superficial operations, as long as a humoral pathology held sway. It is so easy to forget that one could not, in the nature of the case, operate upon the blood or the bile. But as soon as disease was found to lie in local lesions, the extirpation of these became an essential therapeutic process, and surgery moved from an auxiliary position into the very center of medical procedures. This, in turn, attracted able men into the field, led to a demand for improved techniques—of which anesthetics and antiseptics were the most significant—and even transformed the professional position of surgeons themselves.

All this is not intended to detract in any way from the brilliant achievements of both medical bacteriology and surgery in recent generations. These were among the first fruits of modern clinical and pathological research, and there should be no debate about the importance of the two stages of medical advance. One can hardly argue the relative merits of the planting and of the harvest. But in the interest of a sound historical perspective, to say nothing of justice to individuals, one should guard against the popular tendency to proclaim only the harvest and to forget those pioneers who set the seed. Morgagni of Padua, Bichat of Paris, Bright of London, Gerhard of Philadelphia, and Virchow of Berlin will never acquire the popular recognition that is justly accorded to a Pasteur or a Koch, but the former and their professional contemporaries were in all truth the founders of modern medicine.

UNIVERSITY OF PENNSYLVANIA BICENTENNIAL CONFERENCE

Two Centuries of Surgery

By

EVARTS A. GRAHAM, M.D., Sc.D., LL.D.*

THE rapid progress made in all forms of pure and applied science during the last two centuries has been probably as noteworthy in surgery as in any other activity. The changes in both the theory and practice have been so profound that it is difficult for one accustomed to the present-day surgery to imagine the conditions which existed then. In 1740 there was probably no hospital in the whole region of what is now the United States. According to Garrison (1) one had been erected in 1663 on Manhattan Island. Apparently, however, it had ceased to exist by the time this university had its beginning. Not until 1751 was the Pennsylvania Hospital of Philadelphia organized by Dr. Thomas Bond and Benjamin Franklin, to become at any rate the first hospital of this district. It was not until 1756 that it was opened.

Hospitals in those days were merely places to which the poor with serious ailments were taken. They were repulsive to the most hardy and terrifying to the more sensitive. They were infested with vermin, they reeked with foul odors, and the screams of those in pain were often to be heard. The hospital so prevalent now in this country patronized by both rich and poor patients was then unknown. Some of the features, however, of the hospitals of two centuries ago which would seem intolerable to us today probably were much less annoying to those who used them. It was a rougher age and the niceties of physical comfort were far less well developed. For example, odors which would be nauseating to us must have been so commonplace then that they attracted little attention. Bath tubs, plumbing, and sewers were practically unknown even in the European capitals, and probably to even a less degree in a small provincial city like Philadelphia in 1740. Strong perfumes were used by gentlemen

* Bixby Professor of Surgery, School of Medicine, Washington University, St. Louis.

and ladies to counteract the offensive body odors, but the common folk had neither bath tubs nor perfumes. Refuse of all kinds, both physiological and other sorts, was thrown into the unpaved streets, there to be gradually mixed with the mud if it rained or to become dried and blown about if rain did not come. Many an unlucky passer-by was doused with slops from an upstairs window. The virtues of fresh air were not appreciated, and the admission of night air into a house was generally considered positively harmful. Even taxes on windows were in vogue in France and England. From all of this it would seem that the olfactory sense of our forefathers of two centuries ago must have been less reactive to unpleasant odors than is our own; otherwise more effective measures to counteract the universality of stench would have come sooner.

Under the existing circumstances the practice of surgery was exceedingly primitive. Indeed at the time of the founding of this university it was scarcely any more advanced than it had been a century or two previously. It consisted of the practice of external as contrasted with internal medicine. For the most part the operations comprised the opening of abscesses, amputations, and the removal of bladder stones. Operations were not performed to correct conditions within the abdominal or other cavities of the body. Antisepsis and asepsis were unknown; in fact ordinary cleanliness was flouted. In some operating rooms the filth was allowed to accumulate, and even in the nineteenth century in most clinics the "linen dusters" used by the surgeons would be worn continuously for months without a washing and would adorn hooks in the operating room on those days when they were not in use. It is not surprising that inhalations of vinegar and of sulphur were used by the attendants to control the mephitic stench. The pomanders in more common use were hardly strong enough for this job.

Except for whisky and opium there were no anesthetics in common use, and they were usually only for those who could afford to pay for a sufficient amount. Almost never was the patient unconscious when the operation was started. It was hoped by the more compassionate surgeons that he would faint away before it was finished, a hope that was often realized. But the fainting to be effective had to occur early. In lieu of anesthesia the surgeons developed a high degree of speed in operating. He was considered a bungler who would require more than five minutes for most operations. An ordinary thigh amputation was usually completed in three minutes by the able surgeons. Cheselden (2), of London, is said to have performed a lithotomy (the operation for stone in the bladder) in fifty-four seconds. Even a layman can understand that such slap-dash surgery must have been unsatisfactory to the operator. Anyway, now that we have the boon of anesthesia, careful painstaking operating has supplanted the hurried work of a former era. The ability of the surgeon is no longer estimated by the speed with which he performs an operation. Even the great Lister was considered by many contemporaries who lacked a proper judgment to be a poor surgeon because he was unwilling to use the slap-dash methods which had not yet become obsolete. The modern surgeon operates as much with his head as with his hands. The horror of the operation for the unfortunate patient was not over when he was returned to his bed. For it was only in exceptional cases that serious inflammation of the wound did not occur. Indeed patients often died after what today would be considered the most trivial of operations. Most of the deaths of course were due to infections introduced at the time of the operations. But many more must have occurred from the effects of hemorrhage, according to modern standards. More will be said about this point later.

The frequency and seriousness of infections of wounds gave rise to the recognition of the terms "hospital gangrene" and "hospitalism."[1] The plan of erecting hospitals in a series of pavilions became prevalent in the middle and latter part of the nineteenth century because such a plan made it possible either to close off an infected pavilion or to burn it down. It was noticed by many surgeons that patients operated upon at their homes incurred less risk of infection of their wounds than those operated upon in the hospitals.

When one imagines the conditions under which surgery was practised up to the latter part of the last century, until only fifty or sixty years ago, one wonders what kind of men those surgeons could have been. Could anyone but a brute have endured an occupation so characterized by horror, filth, and failure as surgery was two centuries ago? What inducements were there to lead a young man into such a career? Intellectually

[1] Lister's description of the clinical aspects of hospital gangrene is given by Godlee, *Lord Lister,* Macmillan and Co., London, 1918, p. 126.

there was little or no challenge. The diagnosis of those conditions which were treated by the surgeon was for the most part exceedingly simple. In fact, as a general rule, the practice of surgery at the middle of the eighteenth century was not materially different from that of the middle of the sixteenth century, when it was largely in the hands of the barbers. It was chiefly a matter of following the same dogma which was in existence then. Even the increase in the study of anatomy, greatly inspired by the famous Monro dynasty of three generations at Edinburgh, could have been of only academic interest because the operations which were performed were of so simple a sort that very little knowledge of anatomy was required. Could a zeal for helping the most wretched or the promise of great financial returns have been inducements?

To answer these questions one can only say that from what we know now in general of the surgeons of the pre-antiseptic era they seem to have been no different in important characteristics from the majority of the practising surgeons of today. They were probably honest and conscientious, not too good and not too bad, sympathetic to the sufferings of their patients and anxious to do what they could to relieve them. One finds such sympathetic feelings reflected in the writings of many of them. A conspicuous example of tender sympathy of the more or less helpless surgeon to his patient is beautifully presented in Dr. John Brown's *Rab and His Friends*, published in 1859. It is seen also in the case records and notes of operations performed about the year 1800 in the New York Hospital published by Pool and McGowan (3) a few years ago, in which comment is made upon "the fortitude evinced while under the knife." It would be rare to find in the present-day surgical hospital chart any record at all of the manner in which a patient withstood pain. Possibly in the last sixty or seventy-five years more young men with intellectual interests than formerly have become surgeons. Perhaps the almost static condition of surgery for about three centuries was in part due to the fact that it did not appeal to young men with inquisitive minds.

At present, as everyone knows, the horror, the filth, and the failure have been largely removed from the practice of surgery. A patient may now enter an inviting and attractive room in a modern hospital, have his operation with no pain whatever and return to his bed with almost an assurance that, unless he has had an unusually severe condition, he will recover and that he will have a minimum amount of postoperative suffering.

How has this revolutionary change been brought about? As I have already indicated, it is largely the result of what has occurred during the last seventy-five or one hundred years. During the first fifty per cent of the existence of this university, therefore, the surgery that was practised in this city and that was taught in this medical school differed very little from that in vogue in Europe at the time of the discovery of America. During the same time, however, notable progress was made in other activities. In physics practically the whole groundwork of the present-day knowledge of electricity was laid. The discovery of hydrogen and oxygen, of the composition of water, and the formulation of the conception of atomic and molecular weights did much to stimulate the further growth of modern chemistry. In the field of mechanical invention the development of the steam engine, the steamboat, the spinning jenny, and the telegraph were going on at that time. Likewise the American and the French revolutions stimulated new points of view on political philosophy.

It is not strictly accurate to give the impression that no progress whatever was made in surgery from 1740 to 1840. A number of daring new operations were performed, particularly after the year 1800, for the most part successful ligations of arteries for aneurysm which had never before been successfully ligated. But these "trick" operations after all had very little effect on the development of modern surgery. One should, however, take a different attitude towards the courageous and important operation for ovarian cyst performed in 1809 by McDowell in the frontier town of Danville, Kentucky. For the recording of this successful case and of two others in 1817 led to the successful performance of operations on the female pelvis throughout the world even before the antiseptic era. Again the work of the Hunter brothers, William and John, particularly of John, in London, in the latter part of the eighteenth century was of great importance in the struggle against dogma. John Hunter, perhaps more than anyone else, laid the foundations of surgical pathology; that is to say he emphasized the necessity of a clear understanding of disease processes. He had a skeptical inquisitive mind, unwilling to accept dogma and authority. He created a large museum of pathological and anatomical material and he resorted frequently to experimentation in order to arrive at the truth. His general philosophy is expressed in the well-known terse admonition he gave to Jenner, the discoverer of the present-day method of vaccination against smallpox. Jenner discussed his ideas with Hunter and asked his advice. Hunter replied: "Don't think, try; be patient, be accurate." Despite the fact, however, that John Hunter attracted to himself many students who later became famous as a result of his stimulating influence, and despite the very great importance of his work, his influence was not generally felt until a good many years after his death in 1793. Contemporary surgery, therefore, was not markedly affected by him. This may have been due in part to the troubled state of the world, in some respects like the present time. The chaos caused by the French Revolution affected all of Europe, and the American Revolution on this side of the Atlantic had doubtless made American surgeons somewhat less receptive to British influences.

The principal barrier to the development of surgery was the lack of a scientific attitude towards medicine as a whole. Too much respect for authority and dogma necessarily had a blighting influence upon the growth of any new ideas. Much rubbish had to be cleared from the minds of men before there was any room for a spirit which would tolerate a radically different attitude. As an example of the rubbish to which I refer, a quotation from Benjamin Rush, a Philadelphian, the most influential American physician of his time, will serve. Rush was an ardent medical systematist. In other words he belonged to that school of thought which sought to reduce all of the complicated phenomena of disease to a simplified system. In his zeal to excel the other systematists he jumped to a position which nobody could beat. He reduced all of medicine to one disease. No systematist could beat that. No one but Mrs. Eddy could reduce all of medicine to less than one disease. In his "Lectures on the Practice of Physic" (5) in 1796 Rush made this astounding statement to his students:

> I have formerly said that there was but one fever in the world. Be not startled, Gentlemen, follow me and I will say there is but one disease in the world. The proximate cause of disease is irregular convulsive or wrong action in the system affected. This, Gentlemen, is a concise view of my theory of diseases.

Is the balderdash of Mary Baker G. Eddy any worse than that? In my opinion it's not so bad. Moreover, since there was only one disease the treatment of it also was simple. The master systematist, however, could not reduce the treatment to only one form of therapy. No, there were two procedures up the sleeve of every Rush follower. One was bleeding and the other purging. If the patient failed to recover after one bleeding he was bled again and again to reduce the "convulsive action." If death finally intervened it was of course not due to the treatment but in spite of it. There was not much chance for the development of modern surgical conceptions when such nonsense was prevalent. This was a sample of the medicine taught in the medical school of this university almost at the beginning of the nineteenth century; and this medical school by common agreement was the best in America. This nonsense was being uttered by Rush in the very year that Jenner in England successfully protected his first patient against a direct inoculation of smallpox by a previous injection of cowpox. It was nearly twenty years after John Hunter in revolt against dogma had told Jenner: "Don't think, try; be patient, be accurate." Hunter's influence apparently was not felt in America at that time.

It would not have been so bad if Rush's pernicious influence had been confined to Philadelphia, but as an evidence of the almost complete lack of a scientific attitude towards medicine throughout the world at that time, his influence spread not only all over America but to Europe as well. According to Shryock (4), Lettsom (one of the medical great of London at the time) "declared that Rush united 'in an almost unprecedented degree, sagacity and judgment' so that he astonished Europe. And Zimmermann, in Hanover, wrote Lettsom that not only Philadelphia but all humanity should raise a statue to this American prodigy. When Rush died, in 1813, he was widely acclaimed the greatest physician his country had known." However, the ferment of the new attitude brought about by skepticism, resort to the experimental method, and the study of pathology, was working fast. Only thirty years after Rush's death Elisha Bartlett, another prominent American physician, in 1843 (according to Shryock) revaluated his writings thus: "It may be safely said that in the whole vast compass of medical literature, there cannot be found an equal number of pages containing a greater amount and variety of utter nonsense and unqualified absurdity." Lest it might seem that I have a special personal quarrel with Rush I wish to say that I am using him only as an example. As a matter of fact he and I have some common ties. We both were graduated from Princeton and I studied medicine at a medical school in Chicago which bears his name.

I have already indicated that before surgery could advance very rapidly it was necessary that medicine as a whole should be placed upon a more rational basis. The rapid development of the study of anatomy in the eighteenth century did much to bring that about because it popularized a method of studying the body which was a great departure from the former theorizing. In a sense it was an experimental method. When the study of normal anatomy was combined with that of the pathological, as was done by John Hunter, there was no longer any place for the sort of rationalism of Rush and his followers. Rush lived in the twilight of a medical epoch. The dawn of the new day was soon to appear. Garrison (5) has said:

> Up to the time of John Hunter surgery was entirely in French hands, and Paris was the only place where the subject could be properly studied. In Germany, in consequence of the great setback of the Thirty Years' War, general surgery was practiced mainly by the executioner and the barber, or else by the wandering incisors, couchers, and bone-setters, while the army surgeon was called a *Feldscherer*, because it was his duty to shave the officers. . . . In England, Cheselden and Pott were the only two clinical surgeons of first rank before John Hunter's time. The whole period before Hunter was one of enterprise in respect of new amputations, excisions, or other improvements in operative technique, most of which are associated with French names.

Even, however, to those with minds free from the blighting influence of rationalism there still remained certain obstacles to a rapid development of the art of surgery. One of these was the lack of satisfactory anesthesia, another and more important one was the danger of infection inherent in every wound. A third was the lack of understanding of the nature of surgical shock and how to combat it. There were of course many other obstacles which had to be overcome before the highly developed surgery of today could be attained, but when the three which I have mentioned were overcome it became possible for the surgeon safely to carry his art into the hidden parts of the body which previously had defied his attack.

Most laymen see in surgical anesthesia only the humanitarian relief of pain to the patient undergoing an operation. The surgeon, however, sees other benefits also. Anesthesia not only prevents pain but it results in a relaxation of the muscles of the part which is being operated upon. Many of the operations which are performed today would be impossible and others would be more dangerous if the muscles were contracted and rigid as they would be without anesthesia. Moreover, the freedom from pain and the accompanying relaxation which result from anesthesia permit the surgeon to perform a careful, unhurried operation in contrast to the slap-dash performances of the preanesthetic period. A further benefit of anesthesia was that it permitted animal experimentation on a more extensive scale without which the development of the present-day surgery would have been impossible. In spite of the opinions of the antivivisectors, we who perform experiments on animals prefer to avoid suffering, and anyway we prefer to have the animals relaxed for the same reasons as those already mentioned in regard to human patients.

Modern surgical anesthesia originated with the introduction of gases which could be inhaled for the purpose of putting one to sleep. The story of the discovery of anesthesia by that means is one which unfortunately is complicated by many charges and recriminations regarding priority. There seems now, however, to be general agreement that the honor of having been the first to use one of the modern inhalation anesthetics belongs to Crawford Long, who as a country doctor in Jefferson, Georgia, administered ether to a patient to induce anesthesia during an operation for the removal of a small tumor from the back of the neck. This occurred in March 1842, three years after Long received his medical degree from the University of Pennsylvania. A two-cent stamp with his portrait on it was issued by the government last April in honor of the event. Despite the fact that Long seems undoubtedly to have been the first to use ether for surgical anesthesia, he did not publish his discovery and the world was scarcely any better off. William T. G. Morton of Charlton, Massachusetts, a dentist who took up the study of medicine, was told by his preceptor, Dr. Charles T. Jackson, who was also a chemist, of the anesthetic effects of sulphuric ether (the ordinary ether used today for anesthesia). After using that successfully for the extraction of a tooth he persuaded Dr.

John Collins Warren of the Massachusetts General Hospital, Boston, to permit him to induce anesthesia for the removal of a superficial tumor. The operation was performed on October 16, 1846, in what is now called the ether dome of the hospital, before a group of medical visitors. Both the anesthesia and the operation were successful. On November 18 of that year Bigelow (6) in a paper in the *Boston Medical and Surgical Journal*, describing Warren's operation on a patient made unconscious, announced to the world the successful use of surgical anesthesia, entirely ignorant of Crawford Long's earlier work. Instead of "anesthesia," however, he used the word, "insensibility." "Anesthesia" was suggested later by Oliver Wendell Holmes.

Regardless of the question of priority, the discovery of anesthesia was probably America's greatest contribution to surgery. Bigelow's paper apparently electrified the surgical world, and important developments occurred quickly. Articles in the medical journals from various European countries appeared recording experiences with ether anesthesia.

In January of the following year Sir James Y. Simpson, professor of obstetrics at Edinburgh, stirred up a hornet's nest by using ether in some cases of childbirth. This shocking innovation was considered by many to be a direct violation of the Scriptures. Did not God intend that childbirth should be accompanied by travail and pain? The storm abated only when the revered Queen Victoria elected to have anesthesia at the birth of one of her children. The anesthetic used, however, was not ether but chloroform, which had been discovered by the German chemist Liebig, used by Simpson only a few months after he had used ether, and considered by him to be a more desirable anesthetic. For many years afterwards chloroform was the favorite anesthetic drug in Europe, and ether in this country.

Now there are many drugs which are used to produce insensibility for surgical operations, and ether as well as chloroform has become less and less popular. Some of these newer drugs are administered by inhalation and some are injected into a vein. Besides, new methods of anesthesia have been introduced which permit both insensibility and relaxation of the part without the loss of consciousness. One of these methods, called local anesthesia, consists of injecting an anesthetic substance into the tissues surrounding the region to be operated upon or into the nerves leading to the part; another one involves the injection of certain drugs into the spinal canal in order to paralyze temporarily the various nerves below a chosen level.

It is apparent from all this that surgical anesthesia is a rather complicated affair. It involves an accurate knowledge of the effects of each drug on the various organs of the body, the contra-indications for the use of each under certain circumstances, and a high degree of skill in certain procedures such as intravenous and intraspinal injections, in the induction of intratracheal anesthesia and in knowing where to find certain nerves for injection in the performance of regional anesthesia. In addition a knowledge is required of some of the more or less complicated apparatus necessary for the administration of the inhalation anesthetics. It is not surprising therefore that a new word "anesthesiology" has been added recently to our language. There is a new *Journal of Anesthesiology* and an American Board of Anesthesiology for the certification of specialists in that field.

In the early days and until well into the twentieth century the conduction of anesthesia in this country was for the most part very bad, as judged by present-day standards. The methods of administration were crude, but the anesthetists were even cruder. For some curious reason it was felt that any doctor without even any previous training would be competent to administer a safe and satisfactory anesthesia, although scarcely any instruction in the medical schools was given. In nearly all hospitals the internes were the anesthetists, and this highly dangerous and technical procedure was entrusted to the newest and youngest internes at that. The only thing good about that system was that it furnished a good training to the surgeon in the control of his emotions. Many tragedies occurred. The wonder is that there were not more. At the time of which I am speaking the doctors in the United States were not interested in specializing in anesthesia. There were one or two here and there but their number was insignificant. In Europe, however, the situation was different, and many doctors undertook the administration of anesthesia as a career.

Realizing, however, that something must be done to correct the state of affairs in this country, specially selected graduate nurses were given instruction in the art of anesthesia about 1907 by Dr. George Crile and the Mayo brothers. At the Lakeside Hospital, Cleveland, shortly after the first World War a school for the training of nurse anesthetists was created. Since the creation of that school hundreds of nurse anesthetists have been trained not only there but in other similar schools as well. By this movement the administration of anesthetics in most of the hospitals in this country was vastly improved. Permanent professional anesthetists, albeit nurses, were substituted for the incompetent and casual interne ether-pourers. The tremendous benefit to surgery which the nurse anesthetist has been and still is can hardly be overestimated. Because of her even the smaller hospitals can now have reasonably safe and satisfactory anesthesia.

In recent years more and more doctors have taken up anesthesia (or anesthesiology) as a career, and this is as it should be. The American Board of Anesthesiology which awards certificates of proficiency to medical specialists in this field demands as prerequisites a rigorous training of several years and the passing of an examination. Whether or not in time the nurse anesthetist will disappear it is impossible to say. Certainly at present the number of medical anesthetists is grossly inadequate to fill the needs of the country.

The second great obstacle to the progress of surgery during the first century of the existence of this university was, as I have stated earlier, the almost certain development of an infection of the wound with its serious consequences. The surgeon's professional life is spent among wounds. Interest in the healing of wounds has therefore always been a matter of great concern to all thoughtful surgeons. Despite this age-long interest, however, it remained for the Englishman, Joseph Lister, to dispel the common belief that pus is necessary in the healing of a wound and to offer a practical method of eliminating infection from surgical wounds. The detailed story of his work is a fascinating one but too long to recount here. An excellent biography of him was written by his nephew, Sir Rickman J. Godlee (7).

The outstanding features of this epoch-making work are these. Lister's father was much interested in microscopy and made important contributions to the development of the achromatic lens. Joseph was therefore brought up in an atmosphere friendly to a spirit of scientific enquiry. After finishing his medical course at University College, London, in 1853, he went to Edinburgh and immediately as a young house surgeon began a serious and fundamental study of inflammation. In 1860 he received the appointment to the professorship of surgery at Glasgow. In spite of many distracting influences necessitated by his professorship and private practice he continued his active study of problems relating to the infection of wounds. In 1865 his colleague at Glasgow, Dr. Thomas Anderson, the professor of chemistry, called his attention to the papers of Pasteur on fermentation and putrefaction. Lister had already arrived at certain conclusions on infections of wounds, for example (1) putrefaction caused suppuration, and wound infection did not occur without suppuration, (2) that suppuration (decomposition) was in some manner caused by the presence of air, (3) that the gases in the air were not responsible. Pasteur's work apparently supplied the answers to the questions which were bothering Lister. The great Frenchman at that time had demonstrated that putrefaction was fermentation, that dust particles in the air carried microörganisms responsible for the putrefaction and that the air could be freed of those microörganisms by filtration, heat, and in other ways. From all this there developed Lister's idea of sterilizing the air surrounding a field of operation by means of a fine spray of carbolic acid and of applying to the wound dressings saturated with the same substance. Fortunately he achieved successful results. Wounds healed without the development of pus, and of course also the serious constitutional effects of bad wound infections were avoided. Lister gave the name "antisepsis" to his procedures.

Despite his own local successes at Glasgow, however, the methods were only slowly adopted because their significance was not at first appreciated. Many sincere surgeons could not find anything new in the Listerian doctrine. This lack of understanding was due in part to the fact that a French pharmaceutical chemist, Lemaire, in 1863, two years before Lister had heard of Pasteur's work, advocated the use of carbolic acid in surgical wounds. There was, however, a fundamental difference between the conceptions of Lemaire and Lister. The former advised the use of carbolic acid with an idea of arresting a putrefying process in a wound already infected. Lister's idea was to prevent the entrance of bacteria into the wound by the sterilization of the air and by the use of protective dressings saturated with carbolic acid. Every surgeon nowadays knows that the application of an antiseptic drug to the surface of an infected wound accomplishes very little. The exclusion of bacteria is the important thing. Lemaire's conception was therefore unsound and Lister's was correct. Much of the confusion of the contemporary surgeons was due to the fact that Lister used carbolic acid. He did not know of Lemaire's previous recommendation of the use of that antiseptic until later. He selected it because of its success in arresting the putrefaction of sewage which had been achieved at Carlisle. If he had employed some other substance doubtless much of the confusion would have been avoided.

It is interesting that Lister did not know of Lemaire's writings until after many other surgeons knew of them. He was not a scholar and did not know the literature of his subject. Perhaps there is significance in that fact. He might have had less courage to pursue his own ideas.

Gradually there was evolved the present-day method of aseptic surgery. The German von Bergmann played an important rôle in this development. The spray and the extensive use of antiseptic drugs were abandoned. For them was substituted the sterilization by steam of everything that comes into contact with the wound, such as the instruments, the ligatures, the towels, sheets, etc. Sterile rubber gloves, introduced by Halsted of Baltimore, were worn by the surgeon and his assistants, as were also face masks to prevent the passage of bacteria to the wound from the mouth and nose. All of these newer developments were based upon the idea that disease-producing bacteria are not ordinarily present in the air but are transferred from person to person by direct contact, by droplets of nasal or buccal secretions, or by an intermediate agent such as an instrument.

It was now possible to enter any part of the body without pain and with but little danger of infection. Yet an unexpectedly large number of patients died of operations on the table or shortly afterwards. The deaths were generally stated to be due to the vague condition of collapse or shock.

The next important obstacle to overcome in the development of safe surgery, therefore, was the problem of surgical shock. Although even today its nature is not fully understood, yet enough is known about it to enable the surgeon in most cases to prevent a fatal result even when it does occur. It is known, for example, that certain factors have a profound influence in its production. Of these some of the most important are excessive loss of blood, severe injury to the tissues, prolonged deep anesthesia, and chilling of the body. One of the most characteristic clinical features of it is a rapid and severe lowering of the blood pressure because of a disturbance of the circulation of the blood. One of the best ways to restore the blood pressure to a normal level in this condition is to introduce fluid into a vein. Pure water, however, in large quantities is injurious to the blood cells and therefore it cannot be used. Water containing certain concentrations of salts or of glucose can be administered safely, but it passes out of the circulation too rapidly often to be effective in a state of severe shock. The most obvious fluid to use is blood. The idea of blood transfusion after serious hemorrhage was so obvious that it had been used again and again throughout the centuries, occasionally with success but so often with disastrous results that it never gained a place as an important therapeutic procedure until a very short time ago. The explanation of the disasters came with the important discovery by Landsteiner in 1899 that there are different kinds of blood even among human beings. The transfusion of the same type of blood is well tolerated by the recipient, but the use of incompatible blood may have tragic consequences. With this knowledge at hand it is now possible to select a donor who has blood that is compatible with that of the patient. At the present time in at least the best surgical clinics shock has been practically eliminated as a cause of death during or after an operation. Experienced surgeons no longer undertake an operation in which shock is likely to occur without having readily available the necessary means with which to combat it. A compatible blood donor is at hand; or perhaps the blood has already been drawn and with an anticoagulant added to keep it fluid it is in the operating room ready for immediate use. At the beginning of an important operation the surgeon often starts a stream of saline or of glucose solution running slowly into a vein which is continued throughout the operation. In this way shock is often prevented. Moreover, if it occurs from a sudden unexpected hemorrhage blood can be given through the same needle with a minimum of delay. The modern surgeon who does everything possible to prevent the loss of blood looks with horror upon the deliberate copious "blood letting" of Rush and his contemporaries.

In the struggle against surgical shock this country has had

an important part. Dr. George Crile of Cleveland was one of the first to undertake a serious study of it and to emphasize the great importance of taking measures to prevent it. It was largely due to him that nitrous oxide (laughing gas) was introduced into general surgical work as an anesthetic less likely to induce shock than ether or chloroform, which were generally employed at the time. Likewise he was one of the first to utilize blood transfusion in a modern way, to emphasize the importance of the administration of fluid and of the necessity of gentle handling of the tissues, and to devise methods for the avoidance of the chilling of the patient. Surgeons throughout the world became "shock conscious" because of the constant emphasis put upon it by his frequent publications. The simplified technic of transfusion which is now universally employed was the contribution of Lewisohn of New York. Recent important advances by your own Mudd and Ravdin of Philadelphia with their associates have simplified the procedure still more by showing that the plasma of the blood, which is the fluid part of the uncoagulated blood as distinct from the blood cells, can be converted into a powder by a special technic, kept indefinitely in this form and used as a suitable transfusion agent merely by dissolving it in sterile water. With this method it is not even necessary to know to what blood groups the donor and recipient belong. This contribution will be of especially great importance in military surgery.

With the development of anesthesia and the elaboration of aseptic methods it became possible for the first time to extend the application of surgical operations beyond the surface of the body and its extremities to the various internal organs. Before that time operations, sometimes successful, had been performed on some of the organs, notably of the female pelvis, as has been already stated; but the mortality was so high that they were rare occurrences. With a realization, however, that now an internal operation could be performed with a relative assurance of the patient's recovery, surgeons occupied themselves with devising types of operations for the treatment of various diseases affecting the organs. This led also to the invention of many new instruments. In order to compare the efficacy of different operative procedures, particularly in the case of cancer, it became necessary to compile statistics of results with the different types of operations. It is not surprising, therefore, that surgical literature and surgical investigation during the period, roughly, of between 1870 and 1910 were concerned very largely with the development of new operations, new instruments, and statistical studies with reports of extensive series of cases. The Austrian and German surgeons took an early lead in devising operations on the abdominal organs. Of these Billroth, professor of surgery at Vienna from 1867 to 1894, was the most prominent because of his pioneer operations on the stomach, the intestine, and the esophagus. Following his lead many of his brilliant pupils and assistants greatly extended the field of visceral surgery, notably Mikulicz, Czerny, Wölfler, Gersuny, and von Eiselsberg.

The ability to open the abdomen safely in a living subject had a much greater effect than merely to permit the invention of new therapeutic operations. Of at least equal and possibly greater importance were its scientific implications. Previously the knowledge of the pathological changes caused by disease of the abdominal organs had been gained chiefly by autopsy, in other words, by an examination of the patient after the condition was so far advanced as to result in his death. Surgical exploration and removal of the diseased organ afforded an opportunity to study these changes in all stages of their development. This was a fact of such importance that most of our knowledge of the diseases of certain organs has come about through the surgeon's examination or removal of them. As striking examples mention can be made of the gallbladder and duodenum. Naunyn referred to the possibility of surgical examination of living tissues as "autopsies *in vivo*." In later years Moynihan, a great British surgeon, coined the expression "living pathology." But aside from their bearing on the elucidation of pathology, the various operations on the different organs, when transferred to animals and modified to suit the needs of the particular experiment, were of incalculable value in discovering the diverse functions of those organs. For example, it would have been impossible to acquire our present knowledge of the complicated processes of digestion without the use of many of the operations which were and are still employed on the stomach and intestine of the human being. Conversely also, many of the operations introduced into surgical practice had to be performed first on animals in order to study their effects.

It becomes apparent therefore how intimately associated experimental surgery must be with pathology and physiology. All science is interrelated. From what has been said it must be clear that all those who today in the various hospitals are being safely relieved of disabling gallbladders and life-threatening appendices owe a debt of gratitude not only directly to those who devised the operations but more remotely to the French chemist, Pasteur, who interested himself in the study of the fermentation of wine. Even he himself could not possibly have imagined the far-reaching consequences of his fundamental work. One is reminded of Abraham Flexner's expression, "the usefulness of useless knowledge." Who can say what is useless knowledge?

I have paid tribute to Billroth and have mentioned his outstanding pioneer work in the development of abdominal surgery. Of course not all of the important work in that field was done by him. For example, the revelation of the importance of the appendix in the production of peritonitis and death was largely an American contribution in which stand out the names of Fitz of Boston, who from pathological studies showed the significance of the appendix in the condition usually called perityphlitis, and of the surgeons McBurney and Bull of New York, Murphy of Chicago, Deaver of Philadelphia, and others, all of whom elaborated the clinical and diagnostic features and emphasized the necessity of the early removal of a diseased appendix. Dr. W. W. Keen of Philadelphia was a pioneer in the development of the surgery of the gallbladder. The two Mayo brothers and their associates of Rochester, Minnesota, were prominent in developing many new operations on the abdominal organs.

Billroth, however, made another important contribution to surgery, namely the development of a following or a school of surgeons. The most prominent of his disciples have already been named. The idea of a deliberate plan to educate young men to be surgeons comparable to or even to excel the "chief" and to give them the necessary opportunities for their training was new and one that was slow to take root. We are indebted to the German university for it. Most of those whose names have been mentioned in this address left no group of well-trained surgeons behind them. Even the great Lister had no disciples in the sense of which I speak. The idea existed until well into the present century that the surgeon must continuously occupy the limelight of his own group; or at least most surgeons seemed to think so. Little or no encouragement was given to young men to think independently or to grow so tall that they might cast a shadow, however slight, over the effulgent personality of the "chief." The dramatic aspects of surgery were over-emphasized. Operations were performed before admiring throngs in well-filled amphitheatres. No "chief" could permit himself to be eclipsed by an assistant in such an atmosphere. The movement which has done more than anything else to provide a satisfactory education of the young surgeon is the resident system introduced into the Johns Hopkins Hospital by Halsted in the latter part of the last century. In accordance with this plan young men lived in the hospital for a period of years, being allowed more and more independent responsibility in the care of the patients until finally they were permitted to do most of the operating upon the ward patients. At the same time they were encouraged to take up the investigation of problems pertaining to surgery. In principle this system is the best so far devised for the education and training of the surgeon. It combines a suitable amount of supervision by the "chief" with a desirable amount of opportunity for independent growth of the young man. The plan with different modifications has been adopted rather generally in the United States in the university or teaching hospitals. The graded resident system of from four to six years of postgraduate training in surgery under the supervision of a stimulating "chief" assures the fortunate young neophyte of an education which will make him proficient in the science and art of his profession.

The abdomen was the first of the body cavities in which, because of the development of modern surgery, operations became commonplace. Different technics had to be introduced before extensive operations could be performed safely within the cranial cavity, which lodges the brain, and the thorax, in which lie the lungs and heart.

In the development of the surgery of the brain and nervous system American surgeons have had a very conspicuous part. The late Harvey Cushing of Boston occupied in this field a position very similar to that of Billroth in abdominal surgery. In many respects also the men were similar. Both were pioneers and masters of the art of the kind of surgery to which they were important contributors. Both were much interested in medical education and wrote extensively on the subject. Both inspired a group of devoted and distinguished pupils. Many of those in this country and most of those in Europe who now occupy important positions in neurosurgery received at least a part of their training with Cushing. Both had catholic interests extending beyond the field of surgery. Billroth was a musician and wrote a book on harmony. Cushing was an outstanding medical historian, a great bibliophile and the author of a biography of Osler which won a Pulitzer prize. Cushing's contributions to physiology and pathology, however, far outshine those of Billroth. Philadelphians have also contributed greatly to the field of neurosurgery. W. W. Keen operated successfully for a brain tumor (meningioma) in 1888 and tapped the ventricles of the brain for the first time in 1889. The late Charles Frazier of this University was another who contributed greatly to the field. His remarkably brilliant success with his operation for facial neuralgia did much to make the operative relief of this distressing condition a standard procedure. Developments in neurosurgery are progressing so rapidly at the present time that it is difficult for anyone to keep abreast of them.

The thorax presents conditions distinct from the other body cavities which were responsible for making this the last region to be successfully invaded by the surgeon. The pressure within the thorax is less than that of the atmosphere; otherwise the lungs would not be able to expand to take in air. Obviously a large operative incision would result in the creation of an intrathoracic atmosphere pressure which would asphyxiate the patient unless measures were taken to prevent it. Satisfactory technics have now been developed, however, which prevent a serious disturbance of the pressure relationships, and in consequence extensive thoracic surgical operations are now of daily occurrence in some clinics. Many patients with tuberculosis of the lungs who otherwise would die are now restored to normal activity by surgical operations. Those who formerly were hopeless victims of primary cancer of the lung are now offered hope through an operation for the total removal of the affected lung. The first patient upon whom this operation was performed is still living and well nearly eight years later. Smaller portions of lung tissue are now removed for various conditions with a mortality of 5 per cent or less. Even the heart can now be successfully operated upon. Much of the groundwork for the development of thoracic surgery was laid by the German Sauerbruch, but American surgeons have taken it up eagerly and have made the most important contributions to it during the last twenty years. The late Dr. Willy Meyer of New York in 1917 organized the American Association for Thoracic Surgery which, with its annual meeting and its *Journal of Thoracic Surgery,* has become the center of inspiration in this field throughout the world.

Other factors besides the work of the surgeon himself have contributed greatly to the development of modern surgery. One of these is improved methods of diagnosis. There would be no occasion to advise a complicated and perhaps dangerous operation unless the surgeon could be reasonably sure that the patient's symptoms were coming from a diseased condition of the organ about to be attacked. The marvelous assistance rendered by the Roentgen rays (discovered in 1895) has been supplemented by the still more satisfactory evidence of disease obtained by direct vision by means of various instruments such as the cystoscope, the sigmoidoscope, the bronchoscope, etc. These instruments moreover provide a means of obtaining a piece of tissue for microscopic examination, thus often clinching the diagnosis of cancer, for example. In addition, functional tests of various kinds exist which give information about how well a particular organ is working. This information is of value not only in assisting one to indict a particular organ as the one from which the patient's symptoms are coming, but perhaps more important still it often permits the surgeon to estimate whether or not the vital organs are functioning well enough to make the contemplated operation reasonably safe.

Another factor of enormous importance in the development of surgery has been the trained nurse. It is difficult to see how it could have approached its present high level without the sympathetic devotion to the patient and the careful attention to details which can be given only by intelligent women. It is extremely fortunate for us of the present day that Joseph Lister and Florence Nightingale were contemporaries.

Two effects of the development of modern surgery are of interest. One was on the surgeon and the other on the public. The rapid expansion of knowledge of surgical conditions, of operative and of diagnostic technics, resulted in an inevitable specialization. It became impossible for any individual to master the necessary technical requirements and to keep abreast of the rapid progress in anything more than a limited field. Thus various special fields of practice developed. In this respect surgery was not different from practically all other activities. In surgery the specialization has been generally in regard to anatomical regions rather than to diseases. Thus, for example, there are no specialists in streptococcus infections and no specialists in tuberculosis as it affects the whole body. The modern so-called tuberculosis specialist is rather a specialist in diseases of the lungs. At least he usually does not concern himself with tuberculosis of the bones, of the kidneys, etc. One of the earliest specialties was in the diseases of the eye. Since then the number has become almost too numerous to mention. To the layman the word specialist conveys a meaning of one who has superior knowledge. Anyone therefore who designates himself as a specialist is likely to be considered as above the general run of doctors. Moreover there is no legal restraint upon anyone who has a state license to practise from labeling himself as a specialist in any branch of medicine, regardless of how incompetent he may be. In order to protect the public from the unqualified specialist, various examining and qualifying boards of experts have been established under the auspices of the American Medical Association. There is a board for almost every specialty. These boards function without any governmental recognition of any kind, but nevertheless they exert a very definite influence. The names of all those specialists who have received certificates from the boards are designated by appropriate symbols in the directory of the American Medical Association which is available to anybody.

It has been impossible to compress into the space of an hour's lecture any adequate account of the progress which surgery has made during the lifetime of this university. I have been able to dwell, and then too briefly, only on certain outstanding events. I regret that I have not been able to say something about the important influences of the surgical societies and journals, particularly of the American Surgical Association, which was founded here in Philadelphia in 1880 by Samuel Gross, the most outstanding surgeon of his time, and of the influence of the American College of Surgeons. Again the development of the modern hospital, without which modern surgery would be impossible, deserves more than mere mention. I should like also to have had the time to dwell at some length upon the trends of surgical research at the present time as compared with those of even twenty years ago. Finally, I regret that I could not discuss in a manner more satisfactory to me at least the accomplishments which surgery has made in the conquest of disease. It is constantly making more and more inroads into the field of medicine. It is astonishing how many conditions are now amenable to treatment by surgery which even as recently as the time of the Great War were regarded as being strictly non-surgical in character.

The surgeon of today has moved far from his humble position of two centuries ago. There is hardly any contribution to the whole rapidly developing domain of medicine which is unimportant to some field of surgery. The modern surgeon, therefore, if he wishes to be in the forefront of his profession, and particularly if he is ambitious to become a pathfinder, must not only be one well versed in the anatomical, pathological, clinical, and technical foundations of his work, but he will find it necessary even to acquaint himself with distantly related sciences.

REFERENCES

1. Garrison, F. H. An Introduction to the History of Medicine, Philadelphia, 1929, 4th Ed., W. B. Saunders Co., p. 304.
2. Garrison,[1] p. 343.
3. Pool, E. H., and F. J. McGowan. Surgery at the New York Hospital One Hundred Years Ago, New York, 1930, Paul B. Hoeber, Inc.
4. Quoted from Shryock, R. H. The Development of Modern Medicine, Philadelphia, 1936, University of Pennsylvania Press, p. i.
5. Garrison,[1] p. 341.
6. Bigelow, H. J. "Insensibility During Surgical Operations Produced by Inhalation," Boston Med. & Surg. Jour., 35:309, 1846.
7. Godlee, R. J. Lord Lister, London, 1918, The Macmillan Co.

UNIVERSITY OF PENNSYLVANIA
BICENTENNIAL CONFERENCE

Logico-Historical Study of Mechanism, Vitalism, Naturalism

By

EDGAR A. SINGER, PH.D.*

THE literature of our day shows experimental science to be divided between two schools of thought, now generally called Mechanist and Vitalist. The literature of any day these last two thousand years would tell the same tale, but for occasional changes of name. Where an issue dividing scientists is seen to be an experimental issue, it presents no challenge to the philosopher. His interest is limited to the question, How shall we find out? and where all are agreed as to the way of settling a difference of opinion, he can wait with patience to learn the result. But the very history of this world-old and world-wide conflict between schools of experimental science shows that it never has been, is not now, never could become one whose issue turned on the outcome of this or that experiment; nor in all the two thousand years the conflict has lasted, has science been able to come upon any other type of evidence by which the issue might be settled. As one of a class of similar issues, ancient, pervasive, persistent, the manner of whose decision lies still beyond the grasp of experimental science, this Mechanist-Vitalist controversy gives one much to think about.

On more than one occasion, it has seemed to me that to follow a conflict of this class through its history and changes of historic setting brought results not otherwise to be hoped for. But to bring these results, this study of history had to follow a method expressly designed for the purpose. This method, special to the type of problem it is meant to solve, but general to all problems of this type, I have come to call the logico-historical method. The present study may well leave the merits of this method to speak for themselves, if it can offer a satisfactory

* Adam Seybert Professor of Moral and Intellectual Philosophy, University of Pennsylvania.

example of its use in following to some promising outcome one of the most famous, perhaps I should say most notorious, of all the great conflicts of scientific opinion.

Since we are to follow the story of opinions, and differences of opinion, the first step must be to determine what we shall mean by opinions in general, differences of opinion in particular. A few preliminary statements suffice to clarify all understandings, and to put the central problem in its simplest form. First, by *opinion*, let us understand an expression of opinion; and by *expression*, a proposition. This proposition may be as complex as you please in grammatical structure; it may always be written in the form of what the grammarian would call a conjunction of propositions; the logician, a product of factors. In particular, the factors of such a product may always be so chosen as to be themselves unfactorable, and such factors may conveniently be called *prime*.

It will be seen later that all the opinions whose conflict we are to follow through its history are such as can be expressed in the form of a product of prime factors. Not only so, but they may be expressed in terms of a special class of such products; namely, the class that analysts generally qualify by the adjectives *consistent* and *non-redundant*. For the analyst, a product is inconsistent if it can be factored in such a way as to make one factor imply the contradictory of the other; it is redundant, if it can be factored in such a way as to make one factor imply the other. For brevity, I shall throughout the sequel let the simple term, *proposition set*, replace the complex expression, now fully defined, consistent, non-redundant product of prime factors.

Two opinions, each expressed by a proposition set, are incompatible, if and only if the set expressing one implies a proposition whose contradictory is implied by the set expressing the other. When it is understood that the *implication* and *contradiction* here referred to are of the kind called formal, it will be seen that any incompatibility defined in terms of them is also formal, i.e., it is independent of the meaning assigned to any of the terms in which the propositions happen to be expressed. Their incompatibility would be recognized by the formal logician, even if appropriate symbols were substituted for all the terms actually employed. The logico-historical method of following an historic conflict of opinions sets itself the difficult but important task of finding for each opinion entering into the conflict such expression as will reduce every difference of opinion to a formal incompatibility.

To this end, the logico-historical method makes use throughout of what is here called a *classification frame*. Logic itself provides a method by which from any given proposition set a classification frame may be constructed. When, however, this set fulfills not only the conditions general to all such sets, but also a special condition peculiar to what we shall later call *postulate* sets, logic provides a second method of frame construction. It will be convenient to distinguish the two frames to be constructed from a proposition set that is also a postulate set as *proposition frame* and *postulate frame* respectively.

1. *General method of constructing proposition frames from proposition sets.*

i. If P_1P_2 is a proposition set, then $P'_1P_2, P_1P'_2, P'_1P'_2$ are also proposition sets; where, as throughout, P and P′ represent contradictory propositions.

ii. The grammatical disjunction or logical sum $P'_1P_2 + P_1P'_2 + P'_1P'_2$ is the logical contradictory of the product P_1P_2. Generalizing, the sum of any three of these four products is the contradictory of the fourth. Hence, the assemblage of all 4 sets P_1P_2, P'_1P_2, $P_1P'_2$, $P'_1P'_2$ is composed of any one of their number, plus the three sets whose sum is the logical contradictory of that one.

iii. The method by which an assemblage of 4 sets is here formed from a set of 2 factors, may be generalized to form an assemblage of 2^n sets from a set of n factors. We shall call this general method of forming an assemblage of 2^n sets from a set of n factors, the *method of exhaustive contradiction.*

iv. From an assemblage of 2^n sets formed by this method from a given set of n factors a classification of 2^n compartments may be constructed, by assigning to each compartment as its distinguishing set one and only one set of the assemblage. The classification frame so constructed is to be called a *proposition frame.*

2. *Special method of constructing postulate frames from postulate sets.*

i. Def. If a proposition set of n factors implies an (n + 1)th proposition prime in form, while no subset of the set implies this proposition, the n factors (premises) are said to be the *postulates*; the implied proposition (conclusion) is said to be a *theorem* of a deductive system composed of both. The symbol $P_1P_2 \ldots P_n$, P_{n+1} conveniently represents a deductive system, of which $P_1P_2 \ldots P_n$ are the postulates, P_{n+1} a theorem.

ii. If P_1P_2, P_3 is a deductive system, then $P'_3P_2P'_1$ and $P_1P'_3P'_2$ are also deductive systems. The three deductive systems are defined by their respective postulate sets, P_1P_2,P'_3P_2,$P_1P'_3$, of which each set determines the theorem of the deductive system to which it belongs.

iii. The method by which an assemblage of 3 sets is here formed from a postulate set of 2 factors may be generalized to form an assemblage of (n + 1) sets from a postulate set of n factors. We shall call this special method of forming an assemblage of n + 1 sets from a postulate set of n factors, the *method of interchanging contradictories.*

iv. From an assemblage of n + 1 sets formed by this method from a postulate set of n factors, a classification frame of n + 1 compartments may be constructed by assigning to each compartment as its distinguishing set one and only one set of the assemblage. The classification frame so constructed is to be called a *postulate frame.*

So much and no more in the way of logical apparatus is needed by the logico-historical method in any use to which it may be put. The profit its use promises to the historian of science, and it may be through him to the constructive scientist, is best seen in example; and by way of example my choice has fallen on that famous conflict that throughout history has divided Mechanist and Vitalist types of opinion. What the essence of this conflict is; why it should have endured through so long a past; whether it need continue unresolved through all the future, these are the questions on which this study would throw what light it might.

A logico-historical study of any array of historic opinions must begin by setting down a certain conjunction of propositions, or product of propositional factors. The number of these factors, their wording (save for a choice between equivalent wordings) is completely determined by the end the study is to attain. At the outset, one can only say that the analysis to follow will have lacked in rigor, if at the end it is not made clear that no product of factors other than the one chosen could have attained this end. For the present study, the demands of history and of logic are to be met, and are only to be met, by setting down under names whose references will be immediately explained the following conjunction of four propositions:

Democritean (D). Everything in Nature is mechanical in nature.
Aristotelian (A). Some things in Nature are functional in nature.

Common Notion I. Everything mechanical in nature is structural in nature.

Common Notion II. Everything functional in nature is non-structural in nature.

Of the names attached to these propositions, all four have definite historic reference: the first two, D and A, indicate points and dates of origin; the last two, called *common notions* in a somewhat Euclidean sense, are found to be opinions common to all disputants having part in the conflict, and therefore not entering into, but, as we shall show, necessary to a proper understanding of the course of that dispute.

As no three of these propositions implies either the contradictory of the fourth or the fourth itself, their product is neither inconsistent nor redundant; they qualify as a proposition set.

As has been shown, a frame constructed from a proposition set of 4 factors will contain 2^4 or 16 compartments. But of our 4 factors, I and II would not be, as they are said to be, opinions common to all parties to the dispute, were not all compartments whose distinguishing sets included the factors I' or II', empty of historic filling. Hence, any historic opinion that can be defined by a compartment set of our frame, must fall within one of the four compartments whose sets have I and II in common, and whose differentiae are the four products DA, D'A, DA', D'A'. The present study can only set down as subject to the test of documents its own recognition of the historic opinions of which each is defined by one of these four sets, and therefore all are perfectly classified by the frame itself. These schools are

Mechanist (M),	defined	by	the	set	I	II	D	A'
Vitalist (V),	"	"	"	"	I	II	D'	A
Naturalist (N),	"	"	"	"	I	II	D	A
none,	"	"	"	"	I	II	D'	A'

Let us consider now what instruments, not otherwise available, the use of this method of classification puts into the hands of the historian. Two, I think. First, for the comparative historian, it provides a taxonomy of opinions, such that all differences of opinion among the schools classified are determined by the condition stated at the outset; namely, that of formal incompatibility. One sees in fact that of any two compartment sets appearing in this frame each implies (because it contains) a factor whose formal contradictory is implied by (because contained in) the other. Second, for the chronological historian, it provides an exact method of determining critical dates; i.e. the dates at which new opinions are born. For only a formal definition of difference can tell one whether an opinion expressed in the idiom of a given date is or is not different from any that has found its own expression at an earlier date.

Of the taxonomic instrument this particular frame puts at our disposal, full use has already been made in letting it define in a way so completely meeting the demands of logic, the three opinions the history of whose conflict is to be retraced. But, with what results in the way of critical dates will it leave us, if applied as an instrument of chronology? With quite decisive ones. For I think historians would agree that the beginning of the dispute between M and V need not be put much later than Aristotle's time; and that the birth of a well-defined N cannot be put much earlier than Kant's time. Certainly, this new birth can be put no later than some Kantian moment: for Kant's carefully formulated statement of his position in the *K. d. Urtheilskraft* (1790) comes to a conclusion that may be put in this way: No empirical science can reject either the D proposition (as does the V) or the A proposition (as does the M) without flying in the face of all experience. Therefore, the only position on which an experimental science can be built, is one that accepts both D and A (as does the N). And as this is the last of the three schools of science that could be born, our frame has completely served the purpose of the chronologist in telling us when each came into being.

As with this, our frame has answered every possible question as to the what and when of happenings belonging to the history of this famous conflict of scientific opinions, one might suppose the period from Kant to the present day to be void of interest to the historian of this particular dispute. But the logico-historical method, like any other scientific discipline, no sooner wrings from facts the answer to all questions it has thought to ask, than it must needs go over the facts again to see whether they do not suggest questions it had not thought to ask. Now, the Post-Kantian period holds this fact, for whatever it may suggest: the N proposed by Kant won and retained a following among philosophers; among experimental scientists, the very scientists whose demands it was created to meet, its following was negligible or nil. Can one reflect on this bit of history without asking oneself, why? Your habitual user of the logico-historical method cannot. But the very asking brings him to the realization that his method, so far as its possibilities have yet been exploited, has reached its limit of fruitfulness. No proposition frame, from whatever set constructed, can serve to advance us in any way toward an answer to this or any other question: *why?* Its function is entirely performed (and it is a most useful performance) in equipping the historian to define the *what,* and find the *when* of things. If the logico-historical method has anything to contribute to the task of uncovering the *why* of things, it must be by some other device than the use of proposition frames. Can the method furnish such a device? This is what the last third of the present study would consider.

At least, it will be remembered from its first third, that when one can find for purposes of classification a proposition set that is also a postulate set, he has at his disposal not only a general but also a special method of frame construction. But is the set from which the previous proposition frame was constructed also a postulate set? A glance back at it will show it to be. Its factors are the premises of a sorites, they establish a definite conclusion; premises and conclusion together constitute a simple deductive system, having the premises for its postulates, and the conclusion for its theorem. Let us call this theorem, T. From the deductive system I II D A, T the method of interchanging contradictories lets us form a postulate frame of 4 + 1 compartments. Omitting as before the two historically empty compartments whose distinguishing sets would imply factors contradicting one or the other of our common notions (I, II) the remaining compartments of the frame would be distinguished by 3 postulate sets, of which the logician could show each to be the logical equivalent of one of the proposition sets appearing in our proposition frame. Our three schools will then have two quite differently framed but logically equivalent definitions; namely,

		from prop. frame		from post. frame
Mechanist	=	I II D A′	=	I II D T′
Vitalist	=	I II D′ A	=	I II T′ A
Naturalist	=	I II D A	=	I II D A

It will readily be seen that in the second set of definitions as in the first, every difference of opinion is conditioned on the formal incompatibility of the differing opinions. It is, however, not without bearing on the relative competencies of proposition and of postulate defining to account for the *why* of events, to note the difference in type of implication by which incompatibility is established in the two types of definition. Of two proposition sets defining different opinions, each contains at least one factor whose contradictory is contained in the other. Of two postulate sets serving the same function, neither contains a factor whose contradictory is found in the other. Yet for all this apparent harmony, these postulate sets are no less incompatible than were their equivalent proposition sets; for each implies a theorem whose contradictory is a postulate contained in the other. Remembering that our two common notions must be included among the postulates sufficient to establish these incompatibilities, and that such common notions, just because they are not in dispute, are never made explicit, one has occasion to reflect on the part they may play in explaining the psychology of discord. If, indeed, Heraclitus could say, "hidden harmony is better than open"; have we not even more reason to suspect that "hidden discord may be worse than open"? But I anticipate; let me return to the course of a history that will test the point.

The study of historic reasons must begin with that as yet unworded theorem established by the postulates defining the N position. There is, of course, no doubt as to the wording of this theorem: it is the proposition, "Some things non-structural in nature are structural in nature." Certainly our ready-made logic will challenge this conclusion, for is it not the formal contradictory of the proposition, "Nothing non-structural in nature is structural in nature"? And is not this a proposition whose validity is established by formal logic itself? If so, any postulate set that implies (as does the N set) the contradictory of this formally valid proposition must be a logically inconsistent set, and the opinion it defines must be cast into the logician's common receptacle for all inconsistent proposals—the nul class. Certainly we need go no further than this to grasp the logico-historical reason why the empirical science of the past in spite of the strong empirical motive for doing so, has never been able to find a consistent way of fulfilling both the D and A demands.

What then of the future? Evidently, this: *either* the experimental science of the future must agree with the science of the past in accepting the choice between D and A as an inescapable dilemma; *or*, the future will have found some way of escaping the apparent necessity of attributing to the N theorem the form *Some non-a is a*.

But why must any such finding wait for the future? Why has not the thing sought been already found by Kant, who certainly thought he had done all that reason demanded of one who would escape the D - A dilemma (or, as he called it, the D - A antimony)? The answer we are now in position to give is this. Kant discussed the problem in terms of the brief postulate set DA alone. These two propositions are indeed premises of a syllogism establishing a conclusion; but that conclusion is not of a form in which pure logic would recognize a nul proposition. For the only conclusion to be drawn from the premises D and A alone is the proposition, "Some things functional in nature are (also) mechanical in nature." Whatever difficulty there may be in accepting this conclusion, it is not a logical difficulty: this is no nul-proposition. Nevertheless, Kant felt, as all others had felt, that there was difficulty in accepting this conclusion; a difficulty so deep-lying that only the acceptance of proposals reaching to the very root of things could effect its removal. One has only to recall the troubled pages in which Kant struggles to find a solution for his antinomy, to have before his eyes one of the best historic examples of a hidden discord proving itself far worse than any open quarrel could be. Yet the way of formalizing this discord was as open to Kant as it is to us. Had he but continued to the end in the method that so marks the beginning of his critical philosophy, he must in presence of this antinomy have sought and found, among the tacit presuppositions of all who accept either the D or A summing up of experience, the two notions we have set down as *common*. Adding these to the premises of his syllogism, he must have been the first to seize (rigorous logician that he was) the necessity of facing a certain proposition this formal sorites implies: "Nothing can be both functional and mechanical in nature, if it be not both non-structural and structural in nature." Following which he would certainly have agreed, that for no meaning one could possibly give the words *functional* and *mechanical*, could the possession of both a functional and mechanical nature endow the thing possessing them with a gift for enjoying contradictory attributes. Yet just this gift is what any thing both non-s and s in nature must have, if *non-s in nature* and *s in nature* are contradictory terms. But for what other than contradictories can we take these terms? Certainly, the N of 1790, cannot answer a question it did not raise, and no future naturalism can hope to have a following that does not both raise and answer this question. The wonder, then, is not that the empirical scientist, strong as is his motive, can take no comfort in Kantian N. The wonder is that any philosopher can accept Kant as having established a logically permissible hypothesis. And so the date that our proposition frame accepted as marking the birth of a new school of science, our postulate frame can accept for nothing more than the birth-date of a prophet, whom only the future can save from proving a false prophet.

Here, without peering into the future, I should have to leave a study dedicated to the history of the past, had nothing happened between Kant's day and this that might have bearing on the future of N. But something did happen. Something happened which, though not the birth of a new N, was the birth of a new idea without which no new N could be born. For without this idea, the obstacle, now so sharply defined, to the development of a scientific N would stand no better chance in any day than it did in Kant's day, of being removed. This happening was the appearance in 1847 of a *Formal Logic*, the work of the mathematician, Augustus De Morgan.

The new idea this *Logic* offered, was not inspired by any thought of resolving the particular conflict of opinion whose history we have been following, but by a need just as anciently and even more universally felt. It was the need of ending a conflict, not between one thinker and another, but between every thinker and himself. No discourse can be conducted even with oneself without making use of contradictory terms, and yet it seems impossible to find or to invent terms to be used in empirical discourse that fulfil the formal conditions which logic imposes on any pair of terms related as a and non-a. For all logic agrees on two such conditions: (1) Everything is either a or non-a; (2) Nothing is both a and non-a. But take any pair of terms used in familiar discourse, and having the syntactical structure of contradictories; say, *moral* and *immoral*. Do we, or can we use them in a way that conforms with the two conditions set by logic! Do we indeed mean to say that every triangle is either moral or immoral; do we mean to insist that no man can be both? And yet a triangle is something; a man is not nothing. Nor are any other contradictories now in use, nor could any be invented for future use, that would fulfil better than this pair the formal demands of logic. How long, how rich in episode has been the history of the mind's struggle to find a way of discoursing rationally, that is to say, logically with itself! De Morgan's proposal to this end was the simplest and most obvious of all that had ever been made; it was also the most promising. It proposed what might be called a transition from an Absolutist to a Relativist understanding of the sense in which the contradictories of empirical discourse must be made to conform with the conditions imposed upon the a and non-a of formal logic. It is true that *moral* and *immoral* (e.g.) could never be used in meaningful discourse, if *no* kind of thing could be imagined, respecting which all disputants agreed that every thing of this kind must be either moral or immoral, and nothing of this kind could be both. But if all who converse with one another can agree on *some* realm of things within which these conditions hold, then it matters nothing to the purpose of their discourse what things lie outside this realm. Their universe might be as limited as you please, no misunderstanding could arise among them so long as they confined their moral discourse to things within that universe. No doubt, things triangular and non-triangular, which have no place in a moral universe, must be given a place in a geometrical order; no doubt things like you and me which have no place in a strictly moral world must be given a place in some world of looser ethics. But discourse is always rational so long as it knows the universe that its contradictories divide; and so long as it keeps within that universe. In a word, *the* universe of classic logic has been replaced in modern logic by *this* "universe of discourse" and *that*; much as at some prehistoric time *the* horizon must have been replaced by *your* horizon and *mine*.

De Morgan's proposal was promptly seized upon and given many applications, of which our reflections on morality suggest the type. But it will take longer than the scant hundred years that have passed since De Morgan's day, before we shall have plumbed the depths to which the distinction of universes must go, if the entire hierarchy of contradictories needed in scientific discourse is to be exhaustively distributed among them. Nevertheless, I think we have gone far enough in the way of distributing different kinds of thing among different universes of discourse, to draw profit from the result. We may not yet have realized just how many and how different are the kinds of thing we are bound to call by the one name, *thing*; but we have come upon enough distinctions to serve our present purpose. We agree that the thing Hamlet heard behind the arras might have been rat, King or counselor—it must have been a something located in and moving through space-time: *a body*. Again, the thing somebody found unlikable in Dr. Fell might have been envy, hatred, malice or all uncharitableness—it must have been a something common to a whole class of men: *a property*. Once more, the thing the poet found vile in a landscape whose every prospect pleased him might have been mosquitoes, mice or men—it must have been a something composed of many members sharing a common property: *a class*. Of these three things, bodies, properties of bodies, classes of bodies, each one may be classed with things of its own kind; no one can be classed with things of either other kind. But while the things of no one, can be the things of any other of these three universes, yet the classes of one may stand in one to one correspondence with those of another; and because such correspondence must exist between classes of the things called properties and classes of the things called classes, I can put my next question in terms of but two, instead of all three of the universes just distinguished.

Granted that things structural and things non-structural are nothing if not contradictory classes of things, what is the universe of whose things we can say that every one of them is either, and no one is both structural and non-structural? The answer that opens new possibilities to a scientific naturalism is also the one that corresponds with all scientific use: As between bodies and properties, the only things that can fall into the contradictory classes structural and non-structural are properties; and properties are one or the other according as the possession of them by the thing to which they are attributed is or is not independent of that thing's environment. But what, then, is the universe of whose things we say that they have such properties as either are or are not independent of environment, and never both? Surely, only such things as can have environment may belong to this universe; namely, the sort of thing Hamlet heard behind the arras—things located and moving in space-time; bodies. But if this is so, what will be the universe of things non-structural and the things structural in nature, to which the subject and predicate of the N theorems respectively refer? Surely, the first will be the class of bodies possessing properties dependent on environment, i.e. non-structural properties; and the second, the class of bodies possessing properties not dependent on environment, i.e. structural bodies. Is it the same thing to possess contradictory properties, and to possess properties belonging to contradictory classes of properties? We all agree that at a given moment in a gravitational system, no body can have a mass of 1 gr and a mass of 2 gr; or an acceleration of 1 cm per s per s, and acceleration of 2 cm per s per s; but is there any objection to one of these bodies having simultaneously a mass of 1 gr and an acceleration of 1 cm per s per s? Yet in such a system, the mass of a body is, its acceleration is not independent of environment; its mass is a structural, its acceleration a non-structural property.

And now I do not know that I can do better by way of bringing to its time-imposed close, this concrete illustration of the logico-historical method at work, than by translating into the idiom with which the story ends these four colloquially worded propositions with which it began. Its argument must have altogether failed to be convincing, if this translation appear anything other than literal and word for word. For not a word of the original phrasing was wrong; only, the most used term of all was subtly ambiguous. Distinguish the equivocal senses in which the word *thing* was used, and the unequivocal result must be the following, must it not?

D. Every body in Nature has mechanical properties.
I. Every body having mechanical properties has structural properties, i.e., properties independent of environment.
A. Some bodies in Nature have functional properties.
II. Every body having functional properties has non-structural properties; i.e., properties dependent on environment.

Theorem. Some bodies in Nature have both properties that do and properties that do not depend on environment.

Does any one see why they should not?

If not, at least one of the reasons that have kept the past from accepting, as all eperience urged it to accept, both the demand of D and the demand of A has been shown unsound. At least one of the reasons that have sustained the conflict between M and V through all the past need not preserve it through all the future. Kant, the prophet of N, need not turn out to have been a false prophet. Experimental science need not rob itself of Peter, to pay itself with Paul.

UNIVERSITY OF PENNSYLVANIA
BICENTENNIAL CONFERENCE

The Mathematical Way of Thinking

By

HERMANN WEYL, PH.D., DR. ING.*

BY THE mathematical way of thinking I mean first that form of reasoning through which mathematics penetrates into the sciences of the external world—physics, chemistry, biology, economics, etc., and even into our everyday thoughts about human affairs, and secondly that form of reasoning which the mathematician, left to himself, applies in his own field. By the mental process of thinking we try to ascertain truth; it is our mind's effort to bring about its own enlightenment by evidence. Hence, just as truth itself and the experience of evidence, it is something fairly uniform and universal in character. Appealing to the light in our innermost self, it is neither reducible to a set of mechanically applicable rules, nor is it divided into watertight compartments like historic, philosophical, mathematical thinking, etc. We mathematicians are no Ku Klux Klan with a secret ritual of thinking. True, nearer the surface there are certain techniques and differences; for instance, the procedures of fact-finding in a courtroom and in a physical laboratory are conspicuously different. However, you should not expect me to describe the mathematical way of thinking much more clearly than one can describe, say, the democratic way of life.

A movement for the reform of the teaching of mathematics, which some decades ago made quite a stir in Germany under the leadership of the great mathematician Felix Klein, adopted the slogan "functional thinking." The important thing which the average educated man should have learned in his mathematics classes, so the reformers claimed, is thinking in terms of *variables and functions*. A function describes how one variable y depends on another x; or more generally, it maps one variety, the range of a variable element x, upon another (or the same) variety. This idea of function or mapping is certainly one of

* Professor of Mathematics, Institute for Advanced Study.

the most fundamental concepts, which accompanies mathematics at every step in theory and application.

Our federal income tax law defines the tax y to be paid in terms of the income x; it does so in a clumsy enough way by pasting several linear functions together, each valid in another interval or bracket of income. An archeologist who, five thousand years from now, shall unearth some of our income tax returns together with relics of engineering works and mathematical books, will probably date them a couple of centuries earlier, certainly before Galileo and Vieta. Vieta was instrumental in introducing a consistent algebraic symbolism; Galileo discovered the quadratic law of falling bodies, according to which the drop s of a body falling in a vacuum is a quadratic function of the time t elapsed since its release:

$$s = \tfrac{1}{2}gt^2, \qquad (1)$$

g being a constant which has the same value for each body at a given place. By this formula Galileo converted a natural law inherent in the actual motion of bodies into an *a priori* constructed mathematical function, and that is what physics endeavors to accomplish for every phenomenon. The law is of much better design than our tax laws. It has been designed by Nature, who seems to lay her plans with a fine sense for mathematical simplicity and harmony. But then Nature is not, as our income and excess profits tax laws are, hemmed in by having to be comprehensible to our legislators and chambers of commerce.

Right from the beginning we encounter these characteristic features of the mathematical process: 1) variables, like t and s in the formula (1), whose possible values belong to a range, here the range of real numbers, which we can completely survey because it springs from our own free construction, 2) representation of these variables by symbols, and 3) functions or *a priori* constructed mappings of the range of one variable t upon the range of another s. *Time* is the independent variable *kat' exochen.*

In studying a function one should let the independent variable run over its full range. A conjecture about the mutual interdependence of quantities in nature, even before it is checked by experience, may be probed in thought by examining whether it carries through over the whole range of the independent variables. Sometimes certain simple *limiting cases* at once reveal that the conjecture is untenable. Leibnitz taught us by his *principle of continuity* to consider rest not as contradictorily opposed to motion, but as a limiting case of motion. Arguing by continuity he was able *a priori* to refute the laws of impact proposed by Descartes. Ernst Mach gives this prescription: "After having reached an opinion for a special case, one gradually modifies the circumstances of this case as far as possible, and in so doing tries to stick to the original opinion as closely as one can. There is no procedure which leads more safely and with greater mental economy to the simplest interpretation of all natural events." Most of the variables with which we deal in the analysis of nature are continuous variables like time, but although the word seems to suggest it, the mathematical concept is not restricted to this case. The most important example of a discrete variable is given by the sequence of natural numbers or integers 1, 2, 3, . . . Thus the number of divisors of an arbitrary integer n is a function of n.

In Aristotle's logic one passes from the individual to the general by exhibiting certain abstract features in a given object and discarding the remainder, so that two objects fall under the same concept or belong to the same genus if they have those features in common. This descriptive classification, *e.g.*, the description of plants and animals in botany and zoology, is concerned with the actual existing objects. One might say that Aristotle thinks in terms of substance and accident, while the functional idea reigns over the formation of mathematical concepts. Take the notion of ellipse. Any ellipse in the x-y-plane is a set E of points (x, y) defined by a quadratic equation

$$ax^2 + 2bxy + cy^2 = 1$$

whose coefficients a, b, c satisfy the conditions

$$a > 0, c > 0, ac - b^2 > 0.$$

The set E depends on the coefficients a, b, c; we have a function $E(a, b, c)$ which gives rise to an individual ellipse by assigning definite values to the variable coefficients a, b, c. In passing from the individual ellipse to the general notion one does not discard any specific difference, one rather makes certain characteristics (here represented by the coefficients) variable over an *a priori* surveyable range (here described by the inequalities).

The notion thus extends over all *possible*, rather than over all *actually existing*, specifications.[1]

From these preliminary remarks about functional thinking I now turn to a more systematic argument. Mathematics is notorious for the thin air of abstraction in which it moves. This bad reputation is only half deserved. Indeed, the first difficulty the man in the street encounters when he is taught to think mathematically is that he must learn to look things much more squarely in the face; his belief in words must be shattered; he must learn to think more concretely. Only then will he be able to carry out the second step, the step of abstraction where intuitive ideas are replaced by purely symbolic construction.

About a month ago I hiked around Longs Peak in the Rocky Mountain National Park with a boy of twelve, Pete. Looking up at Longs Peak he told me that they had corrected its elevation and that it is now 14,255 feet instead of 14,254 feet last year. I stopped a moment asking myself what this could mean to the boy, and should I try to enlighten him by some Socratic questioning. But I spared Pete the torture, and the comment then withheld will now be served to you. Elevation is elevation above sea level. But there is no sea under Longs Peak. Well, in idea one continues the actual sea level under the solid continents. But how does one construct this ideal closed surface, the geoid, which coincides with the surface of the oceans over part of the globe? If the surface of the ocean were strictly spherical, the answer would be clear. However, nothing of this sort is the case. At this point dynamics comes to our rescue. Dynamically the sea level is a surface of constant potential $\phi = \phi_0$; more exactly ϕ denotes the gravitational potential of the earth, and hence the difference of ϕ at two points P, P' is the work one must put into a small body of mass 1 to transfer it from P to P'. Thus it is most reasonable to define the geoid by the dynamical equation $\phi = \phi_0$. If this constant value of ϕ fixes the elevation zero, it is only natural to define any fixed altitude by a corresponding constant value of ϕ, so that a peak P is called higher than P' if one gains energy by flying from P to P'. The geometric concept of altitude is replaced by the dynamic concept of potential or energy. Even for Pete, the mountain climber, this aspect

[1] Compare about this contrast Ernst Cassirer, *Substanzbegriff und Funktionsbegriff*, 1910, and my critical remark, *Philosophie der Mathematik und Naturwissenschaft*, 1923, p. 111.

is perhaps the most important: the higher the peak the greater—*ceteris paribus*—the mechanical effort in climbing it. By closer scrutiny one finds that in almost every respect the potential is the relevant factor. For instance the barometric measurement of altitude is based on the fact that in an atmosphere of given constant temperature the potential is proportional to the logarithm of the atmospheric pressure, whatever the nature of the gravitational field. Thus atmospheric pressure, generally speaking, indicates potential and not altitude. Nobody who has learned that the earth is round and the vertical direction is not an intrinsic geometric property of space but the direction of gravity should be surprised that he is forced to discard the geometric idea of altitude in favor of the dynamic more concrete idea of potential. Of course there is a relationship to geometry: In a region of space so small that one can consider the force of gravity as constant throughout this region, we have a fixed vertical direction, and potential differences are proportional to differences of altitude measured in that direction. Altitude, height, is a word which has a clear meaning when I ask how high the ceiling of this room is above its floor. The meaning gradually loses precision when we apply it to the relative altitudes of mountains in a wider and wider region. It dangles in the air when we extend it to the whole globe, unless we support it by the dynamical concept of potential. Potential is more concrete than altitude because it is generated by and dependent on the mass distribution of the earth.

Words are dangerous tools. Created for our everyday life they may have their good meanings under familiar limited circumstances, but Pete and the man in the street are inclined to extend them to wider spheres without bothering about whether they then still have a sure foothold in reality. We are witnesses of the disastrous effects of this witchcraft of words in the political sphere where all words have a much vaguer meaning and human passion so often drowns the voice of reason. The scientist must thrust through the fog of abstract words to reach the concrete rock of reality. It seems to me that the science of economics has a particularly hard job, and will still have to spend much effort, to live up to this principle. It is, or should be, common to all sciences, but physicists and mathematicians have been forced to apply it to the most fundamental concepts where the dogmatic resistance is strongest, and thus it has become their second nature. For instance, the first step in explaining relativity theory must always consist in shattering the dogmatic belief in the temporal terms past, present, future. You cannot apply mathematics as long as words still becloud reality.

I return to relativity as an illustration of this first important step preparatory to mathematical analysis, the step guided by the maxim, "Think concretely." As the root of the words *past, present, future,* referring to time, we find something much more tangible than time, namely, the causal structure of the universe. Events are localized in space and time; an event of small extension takes place at a space-time or world point, a here-now. After restricting ourselves to events on a plane E we can depict the events by a graphic timetable in a three-dimensional diagram

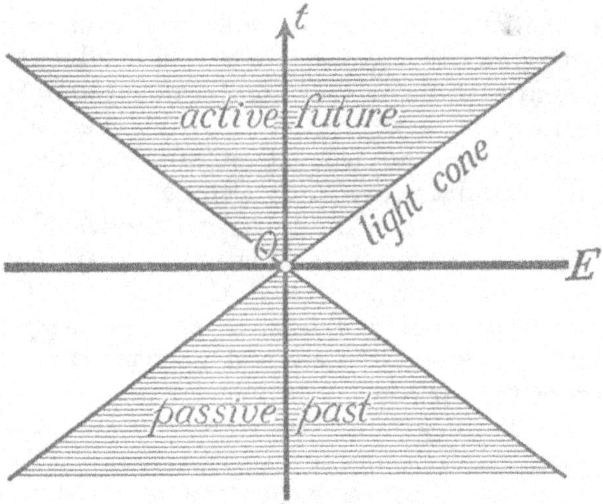

FIG. 1.

with a horizontal E plane and a vertical t-axis on which time t is plotted. A world point is represented by a point in this picture, the motion of a small body by a world line, the propagation of light with its velocity c radiating from a light signal at the world point O by a vertical straight circular cone with vertex at O (light cone). The *active future* of a given world point O, here-now, contains all those events which can still be influenced by what happens at O, while its *passive past* consists of all those world points from which any influence, any message, can reach O. I here-now can no longer change anything that lies outside the active future; all events of which I here-now can have knowledge by direct observation or any records thereof necessarily lie in the passive past. We interpret the words past and future in this causal sense where they express something very real and important, the causal structure of the world.

The new discovery at the basis of the theory of relativity is the fact that no effect may travel faster than light. Hence while we formerly believed that active future and passive past bordered on each other along the cross-section of *present*, the horizontal plane $t =$ const. going through O, Einstein taught us that the active future is bounded by the forward light cone and the passive past by its backward continuation. Active future and passive past are separated by the part of the world lying between these cones, and with this part I am here-now not at all causally connected. The essential positive content of relativity theory is this new insight into the causal structure of the universe. By discussing the various interpretations of such a simple question as whether two men, say Bill on earth and Bob on Sirius, are contemporaries, as to whether it means that Bill can send a message to Bob, or Bob a message to Bill, or even that Bill can communicate with Bob by sending a message and receiving an answer, etc., I often succeed soon in accustoming my listener to thinking in terms of causal rather than his wonted temporal structure. But when I tell him that the causal structure is not a stratification by horizontal layers $t =$ const., but that active future and passive past are of cone-like shape with an interstice between, then some will discern dimly what I am driving at, but every honest listener will say: Now you draw a figure, you speak in pictures; how far does the simile go, and what is the naked truth to be conveyed by it? Our popular writers and news reporters, when they have to deal with physics, indulge in similes of all sorts; the trouble is that they leave the reader helpless in finding out how far these pungent analogies cover the real issue, and therefore more often lead him astray than enlighten him. In our case one has to admit that our diagram is no more than a picture, from which, however, the real thing emerges as soon as we replace the intuitive space in which our diagrams are drawn by its construction in terms of sheer symbols. Then the phrase that the world is a four-dimensional continuum changes from a figurative form of speech into a statement of what is literally true. At this second step the mathematician turns abstract, and here is the point where the layman's understanding most frequently breaks off: the intuitive picture must be exchanged for a symbolic construction. "By its geometric and later by its purely symbolic construction," says Andreas Speiser, "mathematics shook off the fetters of language, and one who knows the enormous work put into this process and its ever recurrent surprising successes can not help feeling that mathematics today is more efficient in its sphere of the intellectual world, than the modern languages in their deplorable state or even music are on their respective fronts." I shall spend most of my time to-day in an attempt to give you an idea of what this magic of symbolic construction is.

To that end I must begin with the simplest, and in a certain sense most profound, example: the natural numbers or *integers* by which we *count* objects. The symbols we use here are strokes put one after another. The objects may disperse, "melt, thaw and resolve themselves into a dew," but we keep this record of their number. What is more, we can by a constructive process decide for two numbers represented through such symbols which one is the larger, namely by checking one against the other, stroke by stroke. This process reveals differences not manifest in direct observation, which in most instances is incapable of distinguishing between even such low numbers as 21 and 22. We are so familiar with these miracles which the number symbols perform that we no longer wonder at them. But this is only the prelude to the mathematical step proper. We do not leave it to chance which numbers we shall meet by counting actual objects, but we generate the open sequence of *all possible* numbers which starts with 1 (or 0) and proceeds by adding to any number symbol n already reached one more stroke, whereby it changes into the following number n'. As I have often said before, being is thus projected onto the background of the possible, or more precisely onto a manifold of possibilities which unfolds by iteration and is open into infinity. Whatever number n we are given, we always deem it possible to pass to the next n'. "Number marches on." This intuition of the "ever one more," of the open countable infinity, is basic for all mathematics. It gives birth to the simplest example of what I termed above an *a priori* surveyable range of variability. According to this process by which the integers are created, functions of an argument ranging over all integers n are to be defined by so-called complete induction, and statements holding for all n are to be proved in the same fashion. The principle of this inference by complete induction is as follows. In order to show that every number n has a certain property V it is sufficient to make sure of two things:

1) 0 has this property;

2) If n is any number which has the property V, then the next number n' has the property V.

It is practically impossible, and would be useless, to write out in strokes the symbol of the number 10^{12}, which the Europeans call a billion and we in this country, a thousand billions. Nevertheless we talk about spending more than 10^{12} cents for our defense program, and the astronomers are still ahead of the financiers. In July the *New Yorker* carried this cartoon: man and wife reading the newspaper over their breakfast and she looking up in puzzled despair: "Andrew, how much *is* seven hundred billion dollars?" A profound and serious question, lady! I wish to point out that only by passing through the *infinite* can we attribute any significance to such figures. 12 as an abbreviation

$$10^{12} = \overset{/\;\;/\;\;/\;\;/\;\;/\;\;/\;\;/\;\;/\;\;/\;\;/\;\;/\;\;/.}{\overline{10\cdot 10\cdot 10\cdot 10\cdot 10\cdot 10\cdot 10\cdot 10\cdot 10\cdot 10\cdot 10\cdot 10}}$$

cannot be understood without defining the function $10 \cdot n$ for *all* n, and this is done through the following definition by complete induction:

$$10 \cdot 0 = 0,$$
$$10 \cdot n' = (10 \cdot n)^{\prime\prime\prime\prime\prime\prime\prime\prime\prime\prime}.$$

The dashes constitute the explicit symbol for 10, and, as previously, each dash indicates transition to the next number. Indian, in particular Buddhist, literature indulges in the possibilities of fixing stupendous numbers by the decimal system of numeration which the Indians invented, *i.e.*, by a combination of sums, products and powers. I mention also Archimedes's treatise "On the counting of sand," and Professor Kasner's Googolplex in his recent popular book on *Mathematics and the Imagination*.

Our conception of *space* is, in a fashion similar to that of natural numbers, depending on a constructive grip on all *possible* places. Let us consider a metallic disk in a plane E. Places on the disk can be marked in *concreto* by scratching little crosses on the plate. But relatively to two axes of coördinates and a standard length scratched into the plate we can also put ideal marks in the plane outside the disk by giving the numerical values of their two coördinates. Each coördinate varies over the *a priori* constructed range of real numbers. In this way astronomy uses our solid earth as a base for plumbing the sidereal spaces. What a marvelous feat of imagination when the Greeks first constructed the shadows which earth and moon, illumined by the sun, cast in empty space and thus explained the eclipses of sun and moon! In analyzing a continuum, like space, we shall here proceed in a somewhat more general manner than by measurement of coördinates and adopt the *topological* viewpoint, so that two continua arising one from the other by continuous deformation are the same to us. Thus the following exposition is at the same time a brief introduction to an important branch of mathematics, topology.

The symbols for the localization of points on the one-dimensional continuum of a straight line are the *real numbers.* I prefer to consider a *closed* one-dimensional continuum, the circle. The most fundamental statement about a continuum is that it may be divided into parts. We catch all the points of a continuum by spanning a net of division over it, which we refine by repetition of a definite process of subdivision *ad infinitum.* Let S be any division of the circle into a number of arcs, say l arcs. From S we derive a new division S' by the process of *normal subdivision,* which consists in breaking each arc into two. The number of arcs in S' will then be $2l$. Running around the circle in a definite sense (orientation) we may distinguish the two pieces, in the order in which we meet them, by the marks 0 and 1; more explicitly, if the arc is denoted by a symbol α then these two pieces are designated as $\alpha 0$ and $\alpha 1$. We start with the division S_0 of the circle into two arcs $+$ and $-$; either is topologically a cell, *i.e.,* equivalent to a segment. We then iterate the process of normal subdivision and thus obtain S_0', S_0'', . . . , seeing to it that the refinement of the division ultimately pulverizes the whole circle. If we had not renounced the use of metric properties we could decree that the normal subdivision takes place by cutting each arc into two *equal* halves. We introduce no such fixation; hence the actual performance of the process involves a wide measure of arbitrariness. However, the *combinatorial scheme* according to which the parts reached at any step border on each other, and according to which the division progresses, is unique and perfectly fixed. Mathematics cares for this symbolic scheme only. By our notation the parts occurring at the consecutive divisions are *catalogued* by symbols of this type

$$+.011010001$$

with + or — before the dot and all following places occupied by either 0 or 1. We see that we arrive at the familiar symbols of binary (not decimal) fractions. A point is caught by an infinite sequence of arcs of the consecutive divisions such that each arc arises from the preceding one by choosing one of the two pieces into which it breaks by the next normal subdivision, and the point is thus fixed by an infinite binary fraction.

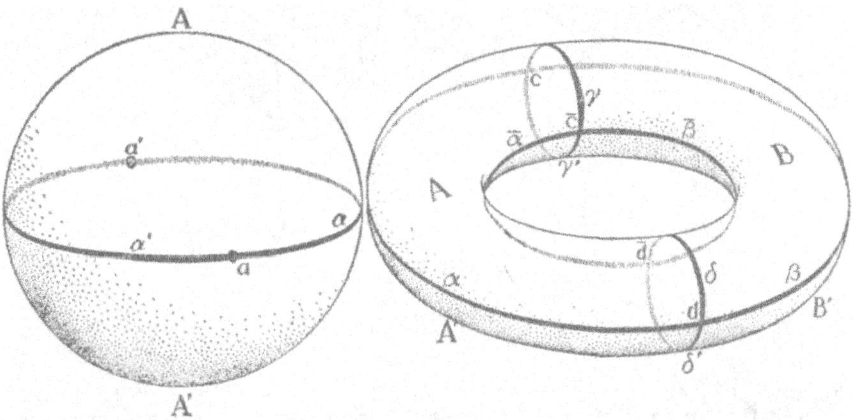

FIGS. 2 AND 3

Let us try to do something similar for two-dimensional continua, *e.g.*, for the surface of a sphere or a torus. The figures show how we may cast a very coarse net over either of them, the one consisting of two, the other of four meshes; the globe is divided into its upper and lower halves by the equator, the torus is welded together from four rectangular plates. The meshes are two-dimensional cells, or briefly, 2-cells which are topologically equivalent to a circular disk. The combinatorial description is facilitated by introducing also the vertices and edges of the division, which are 0- and 1-cells. We attach arbitrary symbols to them and state in symbols for each 2-cell which 1-cells bound it, and for each 1-cell by which 0-cells it is bounded. We then arrive at a *topological scheme* S_0. Here are our two examples:

Sphere. $A \to \alpha, \alpha'$. $A' \to \alpha, \alpha'$. $\alpha \to a, a'$. $\alpha' \to a, a'$.
(→ means: bounded by)

Torus. $A \to \alpha, \bar{\alpha}, \gamma, \delta$. $A' \to \alpha, \bar{\alpha}, \gamma', \delta'$.
$B \to \beta, \bar{\beta}, \gamma, \delta$. $B' \to \beta, \bar{\beta}, \gamma', \delta'$.
$\alpha \to c, d$. $\bar{\alpha} \to \bar{c}, \bar{d}$. $\beta \to c, d$. $\bar{\beta} \to \bar{c}, \bar{d}$.
$\gamma \to c, \bar{c}$. $\gamma' \to c, \bar{c}$. $\delta \to d, \bar{d}$. $\delta' \to d, \bar{d}$.

From this initial stage we proceed by iteration of a universal process of normal subdivision: On each 1-cell $\alpha = ab$ we choose a point which serves as a new vertex α and divides the 1-cell into

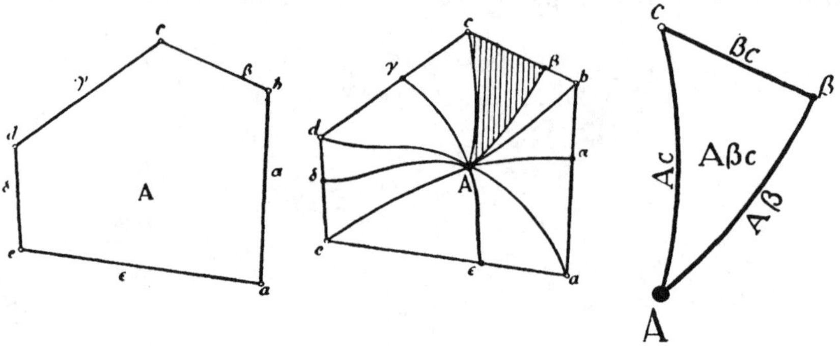

FIG. 4

two segments αa and αb; in each 2-cell A we choose a point A and cut the cell into triangles by joining the newly created vertex A with the old and new vertices on its bounding 1-cells by lines within the 2-cell. Just as in elementary geometry we denote the triangles and their sides by means of their vertices. The figure shows a pentagon before and after subdivision; the triangle $A\beta c$ is bounded by the 1-cells βc, $A\beta$, Ac, the 1-cell Ac for instance by the vertices c and A. We arrive at the following general purely symbolic description of the process by which the subdivided scheme S' is derived from a given topological scheme S. Any symbol $e_2e_1e_0$ made up by the symbols of a 2-cell e_2, a 1-cell e_1 and a 0-cell e_0 in S such that e_2 is bounded by e_1 and e_1 bounded by e_0 represents a 2-cell e'_2 of S'. This 2-cell $e'_2 = e_2e_1e_0$ in S' is part of the 2-cell e_2 in S. The symbols of cells in S' which bound a given cell are derived from its symbol by dropping any one of its constituent letters. Through iteration of this symbolic process the initial scheme S_0 gives rise to a sequence of derived schemes S_0', S_0'', S_0''', . . . What we have done is nothing else than devise a systematic cataloguing of the parts created by consecutive subdivisions. A *point* of our continuum is caught by a sequence

$$e, e', e'', \ldots \qquad (2)$$

which starts with a 2-cell e of S_0 and in which the 2-cell $e^{(n)}$ of the scheme $S^{(n)}$ is followed by one of the 2-cells $e^{(n+1)}$ *of* $S^{(n+1)}$ into which $e^{(n)}$ breaks up by our subdivision. (To do full justice to the inseparability of parts in a continuum this description ought to be slightly altered. But for the present purposes our simplified description will do.) We are convinced that not only may each point be caught by such a sequence (Eudoxos), but that an arbitrarily constructed sequence of this sort always catches a point (Dedekind, Cantor). The fundamental concepts of *limit, convergence* and *continuity* follow in the wake of this construction.

We now come to the decisive step of mathematical abstraction: we forget about what the symbols stand for. The mathematician is concerned with the catalogue alone; he is like the man in the catalogue room who does not care what books or pieces of an intuitively given manifold the symbols of his catalogue denote. He need not be idle; there are many operations which he may carry out with these symbols, without ever having to look at the things they stand for. Thus, replacing the points by their symbols (2) he turns the given manifold into a *symbolic construct* which we shall call the *topological space* $\{S_0\}$ because it is based on the scheme S_0 alone.

The details are not important; what matters is that once the initial finite symbolic scheme S_0 is given we are carried along by an absolutely rigid symbolic construction which leads from S_0 to S_0', from S_0' to S_0'', etc. The idea of iteration, first encountered with the natural numbers, again plays a decisive role. The realization of the symbolic scheme for a given manifold, say a sphere or a torus, as a scheme of consecutive divisions involves a wide margin of arbitrariness restricted only by the requirement that the pattern of the meshes ultimately becomes infinitely fine everywhere. About this point and the closely affiliated requirement that each 2-cell has the topological structure of a circular disk, I must remain a bit vague. However, the mathematician is not concerned with applying the scheme or catalogue to a given manifold, but only with the scheme itself, which contains no haziness whatsoever. And we shall presently see that even the physicist need not care greatly about that application. It was merely for heuristic purposes that we had to go the way from manifold through division to pure symbolism.

In the same purely symbolic way we can evidently construct not only 1- and 2- but also 3, 4, 5, . . . -dimensional manifolds. An n-dimensional scheme S_0 consists of symbols distinguished as 0, 1, 2, . . . , n-cells and associates with each i-cell e_i ($i = 1, 2, . . . , n$) certain ($i - 1$)-cells of which one says that they bound e It is clear how the process of normal subdivision carries over. *A certain such 4-dimensional scheme can be used for the localization of events,* of all possible here-nows; physical quantities which vary in space and time are functions of a variable point ranging over the corresponding symbolically constructed 4-dimensional topological space. In this sense the world *is* a 4-dimensional continuum. The causal structure, of which we talked before, will have to be constructed within the medium of this 4-dimensional world, *i.e.,* out of the symbolic material constituting our topological space. Incidentally the topological viewpoint has been adopted on purpose, because only thus our frame becomes wide enough to embrace both special and general relativity theory. The special theory envisages the causal structure as something geometrical, rigid, given once for all, while in the general theory it becomes flexible and dependent on matter in the same way as, for instance, the electromagnetic field.

In our analysis of nature we reduce the phenomena to simple elements each of which varies over a certain range of possibilities which we can survey *a priori* because we construct these possibilities *a priori* in a purely combinatorial fashion from some purely symbolic material. The manifold of space-time points is one, perhaps the most basic one, of these constructive elements of nature. We dissolve light into plane polarized monochromatic light beams with few variable characteristics like wave length which varies over the symbolically constructed continuum of real numbers. Because of this *a priori* construction we speak of a *quantitative* analysis of nature; I believe the word quantitative, if one can give it a meaning at all, ought to be interpreted in this wide sense. The power of science, as witnessed by the development of modern technology, rests upon the combination of *a priori* symbolic construction with systematic experience in the form of planned and reproducible reactions and their measurements. As material for the *a priori* construction, Galileo and Newton used certain features of reality like space and time which they considered as objective, in opposition to the subjective sense qualities, which they discarded. Hence the important role which geometric figures played in their physics. You probably know Galileo's words in the *Saggiatore* where he says that no one can read the great book of nature "unless he has mastered the code in which it is composed, that is, the mathematical figures and the necessary relations between them." Later we have learned that none of these features of our immediate observation, not even space and time, have a right to survive in a pretended truly objective world, and thus have gradually and ultimately come to adopt a purely symbolic combinatorial construction.

While a set of objects determines its number unambiguously, we have observed that a scheme of division S_0 with its consecutive derivatives S_0', S_0'', . . . can be established on a given manifold in many ways involving a wide margin of arbitrariness. But the question whether two schemes,

$$S_0, S_0', S_0'', \ldots \quad \text{and} \quad T_0, T_0', T_0'', \ldots$$

are fit to describe the same manifold is decidable in a purely mathematical way: it is necessary and sufficient that the two topological spaces $\{S_0\}$ and $\{T_0\}$ can be mapped one upon the other by a continuous one-to-one transformation—a condition which ultimately boils down to a certain relationship called isomorphism between the two schemes S_0 and T_0. (Incidentally the problem of establishing the criterion of isomorphism for two finite schemes in finite combinatorial form is one of the outstanding unsolved mathematical problems.) The connection between a given continuum and its symbolic scheme inevitably carries with it this notion of *isomorphism*; without it and without our understanding that isomorphic schemes are to be considered as not intrinsically different, no more than congruent figures in geometry, the mathematical concept of a topological space would be incomplete. Moreover it will be necessary to formulate precisely the conditions which every topological scheme is required to satisfy. For instance, one such condition demands that each 1-cell be bounded by exactly *two* 0-cells.

I can now say a little more clearly why the physicist is almost as disinterested as the mathematician in the particular way how a certain combinatorial scheme of consecutive divisions is applied to the continuum of here-nows which we called the world. Of course, somehow our theoretical constructions must be put in contact with the observable facts. The historic development of our theories proceeds by heuristic arguments over a long and devious road and in many steps from experience to construction. But systematic exposition should go the other way: first develop the theoretical scheme without attempting to define individually by appropriate measurements the symbols occurring in it as space-time coördinates, electromagnetic field strengths, etc., then describe, as it were in one breath, the contact of the whole system with observable facts. The simplest example I can find is the observed angle between two stars. The symbolic construct in the medium of the 4-dimensional world from which theory determines and predicts the value of this angle includes: (1) the world lines of the two stars, (2) the causal structure of the universe, (3) the world position of the observer and the direction of his world line at the moment of observation. But a continuous deformation, a one-to-one continuous transformation of this whole picture, does not affect the value of the angle. *Isomorphic pictures lead to the same results concerning observable facts.* This is, in its most general form, the *principle of relativity*. The arbitrariness involved in our ascent from the given manifold to the construct is expressed by this principle for the opposite descending procedure, which the systematic exposition should follow.

So far we have endeavored to describe how a mathematical construct is distilled from the given raw material of reality. Let us now look upon these products of distillation with the eye of a pure mathematician. One of them is the sequence of natural numbers and the other the general notion of a topological space $\{ S_0 \}$ into which a topological scheme S_0 develops by consecutive derivations $S_0, S_0', S_0'', \ldots$ In both cases *iteration* is the most decisive feature. Hence all our reasoning must be based on evidence concerning that completely transparent process which generates the natural numbers, rather than on any principles of formal logic like syllogism, etc. The business of the constructive mathematician is *not* to draw logical conclusions. Indeed his arguments and propositions are merely an accompaniment of his actions, his carrying out of constructions. For instance, we run over the sequence of integers 0, 1, 2, . . . by saying alternatingly even, odd, even, odd, etc., and in view of the possibility of this inductive construction which we can extend as far as we ever wish, we formulate the general arithmetical proposition: "Every integer is even or odd." Besides the idea of iteration (or the sequence of integers) we make constant use of mappings or of the functional idea. For instance, just now we have defined a function $\pi(n)$, called parity, with n ranging over all integers and π capable of the two values 0 (even) and 1 (odd), by this induction:

$$\pi(0) = 0;$$
$$\pi(n') = 1 \text{ if } \pi(n) = 0, \quad \pi(n') = 0 \text{ if } \pi(n) = 1.$$

Structures such as the topological schemes are to be studied in the light of the idea of *isomorphism*. For instance, when it comes to introducing operators τ which carry any topological scheme S into a topological scheme $\tau(S)$ one should pay attention only to such operators or functions τ for which isomorphism of S and R entails isomorphism for $\tau(S)$ and $\tau(R)$.

Up to now I have emphasized the constructive character of mathematics. In our actual mathematics there vies with it the non-constructive *axiomatic method*. Euclid's axioms of geometry are the classical prototype. Archimedes employs the method with great acumen and so do later Galileo and Huyghens in erecting the science of mechanics. One defines all concepts in terms of a few undefined basic concepts and deduces all propositions from a number of basic propositions, the axioms, concerning the basic concepts. In earlier times authors were inclined to claim *a priori* evidence for their axioms; however this is an epistemological aspect which does not interest the mathematician. Deduction takes place according to the principles of formal logic, in particular it follows the syllogistic scheme. Such a treatment *more geometrico* was for a long time considered the ideal of every science. Spinoza tried to apply it to ethics. For the mathematician the meaning of the words representing the basic concepts is irrelevant; any interpretation of them which fits, *i.e.*, under which the axioms become true, will be good, and all the propositions of the discipline will hold for such an interpretation because they are all logical consequences of the axioms. Thus n-dimensional Euclidean geometry permits another interpretation where points are distributions of electric current in a given circuit consisting of n branches which connect at certain branch points. For instance, the problem of determining that distribution which results from given electromotoric forces inserted in the various branches of the net corresponds to the geometric construction of orthogonal projection of a point upon a linear subspace. From this standpoint mathematics treats of relations in a hypothetical-deductive manner without binding itself to any particular material interpretation. It is not concerned with the *truth* of axioms, but only with their *consistency*; indeed inconsistency would *a priori* preclude the possibility of our ever coming across a fitting interpretation. "Mathematics is the science which draws necessary conclusions," says B. Peirce in 1870, a definition which was in vogue for decades after. To me it seems that it renders very scanty information about the real nature of mathematics, and you are at present watching my struggle to give a fuller characterization. Past writers on the philosophy of mathematics have so persistently discussed the axiomatic method that I don't think it necessary for me to dwell on it at any greater length, although my exposition thereby becomes somewhat lopsided.

However, I should like to point out that since the axiomatic attitude has ceased to be the pet subject of the methodologists its influence has spread from the roots to all branches of the mathematical tree. We have seen before that topology is to be based on a full enumeration of the axioms which a *topological scheme* has to satisfy. One of the simplest and most basic axiomatic concepts which penetrates all fields of mathematics is that of *group*. Algebra with its "*fields*," "*rings*," etc., is to-day from bottom to top permeated by the axiomatic spirit. Our portrait of mathematics would look a lot less hazy, if time permitted me to explain these mighty words which I have just uttered, group, field and ring. I shall not try it, as little as I have stated the axioms characteristic for a topological scheme. But such notions and their kin have brought it about that modern mathematical research often is a dexterous blending of the constructive and the axiomatic procedures. Perhaps one should be content to note their mutual interlocking. But temptation is great to adopt one of these two views as the genuine primordial way of mathematical thinking, to which the other merely plays a subservient role, and it is possible indeed to carry this standpoint through consistently whether one decides in favor of construction or axiom.

Let us consider the first alternative. Mathematics then consists primarily of construction. The occurring sets of axioms merely *fix the range of variables entering into the construction.* I shall explain this statement a little further by our examples of causal structure and topology. According to the special theory of relativity the causal structure is once for all fixed and can therefore be explicitly constructed. Nay, it is reasonable to construct it together with the topological medium itself, as for instance a circle together with its metric structure is obtained by carrying out the normal subdivision by cutting each arc into two *equal* halves. In the general theory of relativity, however, the causal structure is something flexible; it has only to satisfy certain axioms derived from experience which allow a considerable measure of free play. But the theory goes on by establishing laws of nature which connect the flexible causal structure with other flexible physical entities, distribution of masses, electromagnetic field, etc., and these laws in which the flexible things figure as variables are in their turn *constructed* by the theory in an explicit *a priori* way. Relativistic cosmology asks for the topological structure of the universe as a whole, whether it is open or closed, etc. Of course the topological structure cannot be flexible as the causal struc ture is, but one must have a free outlook on all topological possibilities before one can decide by the testimony of experience which of them is realized by our actual world. To that end one turns to topology. There the topological scheme is bound only by certain axioms; but the topologist derives numerical characters from, or establishes universal connections between, arbitrary topological schemes, and again this is done by explicit construction into which the arbitrary schemes enter as variables. Wherever axioms occur, they ultimately serve to describe the range of variables in explicitly constructed functional relations.

So much about the first alternative. We turn to the opposite view, which subordinates construction to axioms and deduction, and holds that mathematics consists of systems of axioms freely agreed upon, and their necessary conclusions. In a completely axiomatized mathematics construction can come in only secondarily as construction of examples, thus forming the bridge between pure theory and its applications. Sometimes there is only *one* example because the axioms, at least up to arbitrary isomorphisms, determine their object uniquely; then the demand for translating the axiomatic set-up into an explicit construction becomes especially imperative. Much more significant is the remark that an axiomatic system, although it refrains from constructing the mathematical *objects*, constructs the mathematical *propositions* by combined and iterated application of logical rules. Indeed, drawing conclusions from given premises proceeds by certain logical rules which since Aristotle's day one has tried to enumerate completely. Thus on the level of propositions, the axiomatic method is undiluted constructivism. David Hilbert has in our day pursued the axiomatic method to its bitter end where all mathematical propositions, including the axioms, are turned into formulas and the game of deduction proceeds from the axioms by rules which take no account of the meaning of the formulas. The mathematical game is played in silence, without words, like a game of chess. Only the rules have to be explained and communicated in words, and of course any arguing about the possibilities of the game, for instance about its consistency, goes on in the medium of words and appeals to evidence.

If carried so far, the issue between explicit construction and implicit definition by axioms ties up with the last foundations of mathematics. Evidence based on construction refuses to support the principles of Aristotelian logic when these are applied to existential and general propositions in infinite fields like the sequence of integers or a continuum of points. And if the logic of the infinite is taken into account, it seems impossible to axiomatize adequately even the most primitive process, the transition $n \rightarrow n'$ from an integer n to its follower n'. As K. Gödel has shown, there will always be constructively evident arithmetical propositions which can not be deduced from the axioms however you formulate them, while at the same time the axioms, riding roughshod over the subtleties of the constructive infinite, go far beyond what is justifiable by evidence. We are not surprised that a concrete chunk of nature, taken in its isolated phenomenal existence, challenges our analysis by its inexhaustibility and incompleteness; it is for the sake of completeness, as we have seen, that physics projects what is given onto the background of the possible. However, it is surprising that a construct created by mind itself, the sequence of integers, the simplest and most diaphanous thing for the constructive mind, assumes a similar aspect of obscurity and deficiency when viewed from the axiomatic angle. But such is the fact; which casts an uncertain light upon the relationship of evidence and mathematics. In spite, or because, of our deepened critical insight we are to-day less sure than at any previous time of the ultimate foundations on which mathematics rests.

My purpose in this address has not been to show how the inventive mathematical intellect works in its manifold manifestations, in calculus, geometry, algebra, physics, etc., although that would have made a much more attractive picture. Rather, I have attempted to make visible the sources from which all these manifestations spring. I know that in an hour's time I can have succeeded only to a slight degree. While in other fields brief allusions are met by ready understanding, this is unfortunately seldom the case with mathematical ideas. But I should have completely failed if you had not realized at least this much, that mathematics, in spite of its age, is not doomed to progressive sclerosis by its growing complexity, but is still intensely alive, drawing nourishment from its deep roots in mind and nature.

CPSIA information can be obtained
at www.ICGtesting.com
Printed in the USA
BVHW04*0836091018
529575BV00030B/1120/P

9 781512 813807